INDEX

FRACTION

A fraction is a number that represents a part of a whole. A fraction is the quotient of two whole numbers. An example of a fraction is 3/5 in which 3 is the numerator and 5 is the denominator.

Different forms of fractions

There are several forms of fractions:

(1) The absolute value of the numerator of proper fractions is less than the absolute value of their denominator, e.g. 3/4, 2/3, 5/7

(2) The absolute value of the numerator of improper fractions is greater than the absolute value of their denominator, e.g. 5/4, 8/3, 5/3

(3) Mixed fractions have two parts: one part is a whole number and one part is a proper fraction, as

(4) Equivalent fractions have the same amount, such as 1/3 and 2/6

In a/b, if a < b then the fraction is called a proper fraction and if a > b, then the fraction is called an improper fraction. This can be explained in simple language in two ways:

(1) If an amount is divided into b equal parts and a part is taken out of them, then these a parts are called a/b parts of the whole amount, or

(2) If we take such quantities a and divide them into b equal parts, then each is called a/b parts of a quantity. The ratio of two numbers a and b is also expressed by the fraction a/b. If a or b is replaced by any fraction in the fraction a/b, then the fraction so formed is called a mixed fraction, while the original fraction is called a simple fraction, e.g., 3/5 is a simple fraction, but (3/4) / (5/7) are examples of mixed fractions. Mixed fractions can be made more elaborate. Instead of numerator and denominator, a fraction can have a sum, difference, product, quotient of several fractions.

When the denominator of a fraction is a fraction whose denominator is again a fraction and so on, then such a fraction is called a finished fraction.

Rules of fractions

The rules for fractions are as follows:

If the numerator and the denominator are multiplied or divided by the same number, then there is no difference in the value of the fraction, i.e.
a/b = (ak)/(bk)

Addition, subtraction, multiplication and division of fractions:

1. Addition of Fractions

2/3 + 2/3

Before adding any fractions, we need to find out whether they have the same denominator or not. We can see above that both of these fractions have the same denominator.

2. We saw that both the fractions have the same denominator, so we will add the numerators keeping the denominator the same.

2+2 = 4

3. So now we will add these fractions as follows:

2/3 + 2/3 = 4/3

As you can see in the answer only the fractions have been added but the numerator remains the same.

2. Subtraction of fraction

6/7 – 2/7

As you can see in the above fraction, the denominators of both the fractions are equal. We won't have much difficulty in subtracting the fraction.
We first write down the denominator. As we know that in the coming fraction the denominator is going to be 7 only.
6/7 – 2/7 = –/7

Now we have to subtract the numerators from each other.
6-2 = 4

Solution of this fraction will be : 4/7

Thus we can subtract any two fractions that have the same denominator.

3. Multiplication of fractions

3/5*6/7

As you can see there are two fractions given above which are both in their simplest form. We cannot write them in any simpler form.

Now to multiply fractions, we multiply the denominator of the first fraction by the denominator of the second fraction.

5*7 = 35

As we can see that the denominator of the following fraction will be 35. Now we will multiply the numerators of both the fractions to find the numerator of the coming fraction.

3*6 = 18

We can see that now we have found the numerator as well. Now we will write the numerator and the denominator on top of each other. Therefore

The solution for multiplying these fractions is: 18/35

4. Division of fractions

2/3 / 5/7

As you can see above we have two fractions that we need to divide. As we know that in order to divide a fraction, first of all we have to write the numerator of the second fraction in place of the denominator and the denominator in place of the numerator. This will invert the second fraction and the sign of division will change to the sign of multiplication.

2/3 * 7/5

As you can see in the above given fractions, here now the sign of division has come in place of multiplication. So now we will multiply the numerator of the first fraction by the numerator of the second fraction and multiply the denominator of the first fraction by the denominator of the second fraction to get the solution.

2*7 = 14, 3*5 = 15

On multiplying numerator and denominator, we get a new fraction which is as follows:

14 / 15

As we can see above given fraction is in its simplest form so now we cannot write it in simplest form so this will be the solution of these fractions.

Solution to Division of Fractions: 14/15

ADDITION OF FRACTIONS

1). $\dfrac{2}{9} + \dfrac{5}{9} =$

2). $\dfrac{3}{10} + \dfrac{5}{10} =$

3). $\dfrac{1}{9} + \dfrac{5}{9} =$

4). $\dfrac{1}{10} + \dfrac{4}{10} =$

5). $\dfrac{2}{7} + \dfrac{2}{7} =$

6). $\dfrac{1}{6} + \dfrac{3}{6} =$

7). $\dfrac{1}{3} + \dfrac{1}{3} =$

8). $\dfrac{3}{8} + \dfrac{3}{8} =$

9). $\dfrac{1}{4} + \dfrac{1}{4} =$

10). $\dfrac{2}{11} + \dfrac{4}{11} =$

11). $\dfrac{1}{3} + \dfrac{1}{3} =$

12). $\dfrac{1}{12} + \dfrac{5}{12} =$

13). $\dfrac{2}{5} + \dfrac{2}{5} =$

14). $\dfrac{1}{10} + \dfrac{3}{10} =$

15). $\dfrac{1}{8} + \dfrac{2}{8} =$

16). $\dfrac{1}{11} + \dfrac{1}{11} =$

17). $\dfrac{1}{4} + \dfrac{2}{4} =$

18). $\dfrac{1}{11} + \dfrac{1}{11} =$

19). $\dfrac{1}{9} + \dfrac{6}{9} =$

20). $\dfrac{1}{12} + \dfrac{9}{12} =$

21). $\frac{2}{6} + \frac{3}{6} =$

22). $\frac{2}{10} + \frac{3}{10} =$

23). $\frac{1}{11} + \frac{7}{11} =$

24). $\frac{1}{8} + \frac{2}{8} =$

25). $\frac{3}{12} + \frac{8}{12} =$

26). $\frac{2}{7} + \frac{2}{7} =$

27). $\frac{3}{12} + \frac{6}{12} =$

28). $\frac{1}{3} + \frac{1}{3} =$

29). $\frac{2}{12} + \frac{2}{12} =$

30). $\frac{1}{5} + \frac{3}{5} =$

31). $\frac{1}{3} + \frac{1}{3} =$

32). $\frac{2}{10} + \frac{6}{10} =$

33). $\frac{1}{12} + \frac{5}{12} =$

34). $\frac{4}{12} + \frac{5}{12} =$

35). $\frac{1}{9} + \frac{3}{9} =$

36). $\frac{1}{6} + \frac{1}{6} =$

37). $\frac{1}{4} + \frac{2}{4} =$

38). $\frac{4}{11} + \frac{6}{11} =$

39). $\frac{1}{8} + \frac{4}{8} =$

40). $\frac{1}{5} + \frac{1}{5} =$

41). $\dfrac{1}{12} + \dfrac{4}{12} =$

42). $\dfrac{1}{3} + \dfrac{1}{3} =$

43). $\dfrac{2}{9} + \dfrac{3}{9} =$

44). $\dfrac{2}{7} + \dfrac{2}{7} =$

45). $\dfrac{1}{12} + \dfrac{9}{12} =$

46). $\dfrac{1}{9} + \dfrac{2}{9} =$

47). $\dfrac{1}{11} + \dfrac{6}{11} =$

48). $\dfrac{1}{8} + \dfrac{5}{8} =$

49). $\dfrac{3}{10} + \dfrac{6}{10} =$

50). $\dfrac{2}{10} + \dfrac{7}{10} =$

51). $\dfrac{5}{12} + \dfrac{6}{12} =$

52). $\dfrac{1}{10} + \dfrac{7}{10} =$

53). $\dfrac{1}{4} + \dfrac{2}{4} =$

54). $\dfrac{1}{9} + \dfrac{5}{9} =$

55). $\dfrac{1}{12} + \dfrac{4}{12} =$

56). $\dfrac{1}{11} + \dfrac{3}{11} =$

57). $\dfrac{1}{7} + \dfrac{1}{7} =$

58). $\dfrac{1}{6} + \dfrac{3}{6} =$

59). $\dfrac{1}{11} + \dfrac{4}{11} =$

60). $\dfrac{1}{10} + \dfrac{4}{10} =$

61). $\frac{2}{3} + \frac{3}{5} =$ 71). $\frac{2}{3} + \frac{2}{10} =$

62). $\frac{2}{4} + \frac{1}{2} =$ 72). $\frac{7}{11} + \frac{2}{22} =$

63). $\frac{7}{10} + \frac{3}{4} =$ 73). $\frac{4}{8} + \frac{10}{12} =$

64). $\frac{2}{3} + \frac{1}{2} =$ 74). $\frac{3}{7} + \frac{5}{14} =$

65). $\frac{5}{10} + \frac{1}{2} =$ 75). $\frac{6}{10} + \frac{2}{3} =$

66). $\frac{6}{10} + \frac{1}{4} =$ 76). $\frac{1}{5} + \frac{3}{4} =$

67). $\frac{3}{4} + \frac{7}{10} =$ 77). $\frac{8}{27} + \frac{1}{9} =$

68). $\frac{1}{2} + \frac{3}{5} =$ 78). $\frac{1}{6} + \frac{8}{9} =$

69). $\frac{3}{4} + \frac{4}{5} =$ 79). $\frac{2}{11} + \frac{11}{22} =$

70). $\frac{2}{4} + \frac{1}{10} =$ 80). $\frac{2}{14} + \frac{1}{7} =$

81). $\dfrac{1}{13} + \dfrac{2}{4} =$

82). $\dfrac{9}{30} + \dfrac{3}{60} =$

83). $\dfrac{8}{60} + \dfrac{3}{12} =$

84). $\dfrac{13}{14} + \dfrac{6}{7} =$

85). $\dfrac{2}{16} + \dfrac{6}{8} =$

86). $\dfrac{11}{13} + \dfrac{4}{26} =$

87). $\dfrac{16}{46} + \dfrac{13}{23} =$

88). $\dfrac{10}{11} + \dfrac{4}{55} =$

89). $\dfrac{14}{32} + \dfrac{2}{4} =$

90). $\dfrac{9}{52} + \dfrac{4}{26} =$

91). $\dfrac{2}{20} + \dfrac{3}{5} =$

92). $\dfrac{9}{10} + \dfrac{5}{20} =$

93). $\dfrac{3}{9} + \dfrac{17}{45} =$

94). $\dfrac{1}{4} + \dfrac{17}{34} =$

95). $\dfrac{1}{16} + \dfrac{2}{4} =$

96). $\dfrac{3}{4} + \dfrac{16}{76} =$

97). $\dfrac{1}{6} + \dfrac{15}{24} =$

98). $\dfrac{14}{21} + \dfrac{3}{7} =$

99). $\dfrac{1}{4} + \dfrac{3}{5} =$

100). $\dfrac{19}{38} + \dfrac{13}{19} =$

ANSWER

1). $\dfrac{2}{9} + \dfrac{5}{9} = \dfrac{7}{9}$

2). $\dfrac{3}{10} + \dfrac{5}{10} = \dfrac{8}{10}$

3). $\dfrac{1}{9} + \dfrac{5}{9} = \dfrac{6}{9}$

4). $\dfrac{1}{10} + \dfrac{4}{10} = \dfrac{5}{10}$

5). $\dfrac{2}{7} + \dfrac{2}{7} = \dfrac{4}{7}$

6). $\dfrac{1}{6} + \dfrac{3}{6} = \dfrac{4}{6}$

7). $\dfrac{1}{3} + \dfrac{1}{3} = \dfrac{2}{3}$

8). $\dfrac{3}{8} + \dfrac{3}{8} = \dfrac{6}{8}$

9). $\dfrac{1}{4} + \dfrac{1}{4} = \dfrac{2}{4}$

10). $\dfrac{2}{11} + \dfrac{4}{11} = \dfrac{6}{11}$

11). $\dfrac{1}{3} + \dfrac{1}{3} = \dfrac{2}{3}$

12). $\dfrac{1}{12} + \dfrac{5}{12} = \dfrac{6}{12}$

13). $\dfrac{2}{5} + \dfrac{2}{5} = \dfrac{4}{5}$

14). $\dfrac{1}{10} + \dfrac{3}{10} = \dfrac{4}{10}$

15). $\dfrac{1}{8} + \dfrac{2}{8} = \dfrac{3}{8}$

16). $\dfrac{1}{11} + \dfrac{1}{11} = \dfrac{2}{11}$

17). $\dfrac{1}{4} + \dfrac{2}{4} = \dfrac{3}{4}$

18). $\dfrac{1}{11} + \dfrac{1}{11} = \dfrac{2}{11}$

19). $\dfrac{1}{9} + \dfrac{6}{9} = \dfrac{7}{9}$

20). $\dfrac{1}{12} + \dfrac{9}{12} = \dfrac{10}{12}$

21). $\dfrac{2}{6} + \dfrac{3}{6} = \quad \dfrac{5}{6}$

22). $\dfrac{2}{10} + \dfrac{3}{10} = \quad \dfrac{5}{10}$

23). $\dfrac{1}{11} + \dfrac{7}{11} = \quad \dfrac{8}{11}$

24). $\dfrac{1}{8} + \dfrac{2}{8} = \quad \dfrac{3}{8}$

25). $\dfrac{3}{12} + \dfrac{8}{12} = \quad \dfrac{11}{12}$

26). $\dfrac{2}{7} + \dfrac{2}{7} = \quad \dfrac{4}{7}$

27). $\dfrac{3}{12} + \dfrac{6}{12} = \quad \dfrac{9}{12}$

28). $\dfrac{1}{3} + \dfrac{1}{3} = \quad \dfrac{2}{3}$

29). $\dfrac{2}{12} + \dfrac{2}{12} = \quad \dfrac{4}{12}$

30). $\dfrac{1}{5} + \dfrac{3}{5} = \quad \dfrac{4}{5}$

31). $\dfrac{1}{3} + \dfrac{1}{3} = \quad \dfrac{2}{3}$

32). $\dfrac{2}{10} + \dfrac{6}{10} = \quad \dfrac{8}{10}$

33). $\dfrac{1}{12} + \dfrac{5}{12} = \quad \dfrac{6}{12}$

34). $\dfrac{4}{12} + \dfrac{5}{12} = \quad \dfrac{9}{12}$

35). $\dfrac{1}{9} + \dfrac{3}{9} = \quad \dfrac{4}{9}$

36). $\dfrac{1}{6} + \dfrac{1}{6} = \quad \dfrac{2}{6}$

37). $\dfrac{1}{4} + \dfrac{2}{4} = \quad \dfrac{3}{4}$

38). $\dfrac{4}{11} + \dfrac{6}{11} = \quad \dfrac{10}{11}$

39). $\dfrac{1}{8} + \dfrac{4}{8} = \quad \dfrac{5}{8}$

40). $\dfrac{1}{5} + \dfrac{1}{5} = \quad \dfrac{2}{5}$

41). $\dfrac{1}{12} + \dfrac{4}{12} =$ $\dfrac{5}{12}$

51). $\dfrac{5}{12} + \dfrac{6}{12} =$ $\dfrac{11}{12}$

42). $\dfrac{1}{3} + \dfrac{1}{3} =$ $\dfrac{2}{3}$

52). $\dfrac{1}{10} + \dfrac{7}{10} =$ $\dfrac{8}{10}$

43). $\dfrac{2}{9} + \dfrac{3}{9} =$ $\dfrac{5}{9}$

53). $\dfrac{1}{4} + \dfrac{2}{4} =$ $\dfrac{3}{4}$

44). $\dfrac{2}{7} + \dfrac{2}{7} =$ $\dfrac{4}{7}$

54). $\dfrac{1}{9} + \dfrac{5}{9} =$ $\dfrac{6}{9}$

45). $\dfrac{1}{12} + \dfrac{9}{12} =$ $\dfrac{10}{12}$

55). $\dfrac{1}{12} + \dfrac{4}{12} =$ $\dfrac{5}{12}$

46). $\dfrac{1}{9} + \dfrac{2}{9} =$ $\dfrac{3}{9}$

56). $\dfrac{1}{11} + \dfrac{3}{11} =$ $\dfrac{4}{11}$

47). $\dfrac{1}{11} + \dfrac{6}{11} =$ $\dfrac{7}{11}$

57). $\dfrac{1}{7} + \dfrac{1}{7} =$ $\dfrac{2}{7}$

48). $\dfrac{1}{8} + \dfrac{5}{8} =$ $\dfrac{6}{8}$

58). $\dfrac{1}{6} + \dfrac{3}{6} =$ $\dfrac{4}{6}$

49). $\dfrac{3}{10} + \dfrac{6}{10} =$ $\dfrac{9}{10}$

59). $\dfrac{1}{11} + \dfrac{4}{11} =$ $\dfrac{5}{11}$

50). $\dfrac{2}{10} + \dfrac{7}{10} =$ $\dfrac{9}{10}$

60). $\dfrac{1}{10} + \dfrac{4}{10} =$ $\dfrac{5}{10}$

61). $\dfrac{2}{3} + \dfrac{3}{5} = \dfrac{19}{15}$

62). $\dfrac{2}{4} + \dfrac{1}{2} = \dfrac{4}{4}$

63). $\dfrac{7}{10} + \dfrac{3}{4} = \dfrac{29}{20}$

64). $\dfrac{2}{3} + \dfrac{1}{2} = \dfrac{7}{6}$

65). $\dfrac{5}{10} + \dfrac{1}{2} = \dfrac{10}{10}$

66). $\dfrac{6}{10} + \dfrac{1}{4} = \dfrac{17}{20}$

67). $\dfrac{3}{4} + \dfrac{7}{10} = \dfrac{29}{20}$

68). $\dfrac{1}{2} + \dfrac{3}{5} = \dfrac{11}{10}$

69). $\dfrac{3}{4} + \dfrac{4}{5} = \dfrac{31}{20}$

70). $\dfrac{2}{4} + \dfrac{1}{10} = \dfrac{12}{20}$

71). $\dfrac{2}{3} + \dfrac{2}{10} = \dfrac{26}{30}$

72). $\dfrac{7}{11} + \dfrac{2}{22} = \dfrac{16}{22}$

73). $\dfrac{4}{8} + \dfrac{10}{12} = \dfrac{32}{24}$

74). $\dfrac{3}{7} + \dfrac{5}{14} = \dfrac{11}{14}$

75). $\dfrac{6}{10} + \dfrac{2}{3} = \dfrac{38}{30}$

76). $\dfrac{1}{5} + \dfrac{3}{4} = \dfrac{19}{20}$

77). $\dfrac{8}{27} + \dfrac{1}{9} = \dfrac{11}{27}$

78). $\dfrac{1}{6} + \dfrac{8}{9} = \dfrac{19}{18}$

79). $\dfrac{2}{11} + \dfrac{11}{22} = \dfrac{15}{22}$

80). $\dfrac{2}{14} + \dfrac{1}{7} = \dfrac{4}{14}$

81). $\frac{1}{13} + \frac{2}{4} = \frac{30}{52}$

82). $\frac{9}{30} + \frac{3}{60} = \frac{21}{60}$

83). $\frac{8}{60} + \frac{3}{12} = \frac{23}{60}$

84). $\frac{13}{14} + \frac{6}{7} = \frac{25}{14}$

85). $\frac{2}{16} + \frac{6}{8} = \frac{14}{16}$

86). $\frac{11}{13} + \frac{4}{26} = \frac{26}{26}$

87). $\frac{16}{46} + \frac{13}{23} = \frac{42}{46}$

88). $\frac{10}{11} + \frac{4}{55} = \frac{54}{55}$

89). $\frac{14}{32} + \frac{2}{4} = \frac{30}{32}$

90). $\frac{9}{52} + \frac{4}{26} = \frac{17}{52}$

91). $\frac{2}{20} + \frac{3}{5} = \frac{14}{20}$

92). $\frac{9}{10} + \frac{5}{20} = \frac{23}{20}$

93). $\frac{3}{9} + \frac{17}{45} = \frac{32}{45}$

94). $\frac{1}{4} + \frac{17}{34} = \frac{51}{68}$

95). $\frac{1}{16} + \frac{2}{4} = \frac{9}{16}$

96). $\frac{3}{4} + \frac{16}{76} = \frac{73}{76}$

97). $\frac{1}{6} + \frac{15}{24} = \frac{19}{24}$

98). $\frac{14}{21} + \frac{3}{7} = \frac{23}{21}$

99). $\frac{1}{4} + \frac{3}{5} = \frac{17}{20}$

100). $\frac{19}{38} + \frac{13}{19} = \frac{45}{38}$

ADDING THREE FRACTIONS

1). $\dfrac{5}{7} + \dfrac{8}{7} + \dfrac{5}{7} =$

2). $\dfrac{7}{7} + \dfrac{4}{7} + \dfrac{4}{7} =$

3). $\dfrac{3}{5} + \dfrac{7}{5} + \dfrac{4}{5} =$

4). $\dfrac{1}{12} + \dfrac{4}{12} + \dfrac{1}{12} =$

5). $\dfrac{1}{6} + \dfrac{2}{6} + \dfrac{2}{6} =$

6). $\dfrac{3}{7} + \dfrac{3}{7} + \dfrac{1}{7} =$

7). $\dfrac{2}{7} + \dfrac{6}{7} + \dfrac{2}{7} =$

8). $\dfrac{7}{9} + \dfrac{4}{9} + \dfrac{6}{9} =$

9). $\dfrac{1}{7} + \dfrac{7}{7} + \dfrac{2}{7} =$

10). $\dfrac{9}{8} + \dfrac{3}{8} + \dfrac{2}{8} =$

11). $\dfrac{5}{8} + \dfrac{4}{8} + \dfrac{7}{8} =$

12). $\dfrac{1}{9} + \dfrac{3}{9} + \dfrac{7}{9} =$

13). $\dfrac{2}{8} + \dfrac{4}{8} + \dfrac{6}{8} =$

14). $\dfrac{6}{7} + \dfrac{6}{7} + \dfrac{6}{7} =$

15). $\dfrac{4}{11} + \dfrac{2}{11} + \dfrac{6}{11} =$

16). $\dfrac{2}{9} + \dfrac{8}{9} + \dfrac{7}{9} =$

17). $\dfrac{5}{8} + \dfrac{3}{8} + \dfrac{6}{8} =$

18). $\dfrac{5}{4} + \dfrac{4}{4} + \dfrac{3}{4} =$

19). $\dfrac{7}{11} + \dfrac{3}{11} + \dfrac{1}{11} =$

20). $\dfrac{6}{6} + \dfrac{3}{6} + \dfrac{1}{6} =$

21). $\frac{1}{4} + \frac{9}{10} + \frac{1}{5} =$

22). $\frac{2}{5} + \frac{1}{3} + \frac{1}{2} =$

23). $\frac{1}{3} + \frac{1}{2} + \frac{2}{5} =$

24). $\frac{2}{4} + \frac{2}{3} + \frac{1}{2} =$

25). $\frac{1}{2} + \frac{3}{5} + \frac{2}{3} =$

26). $\frac{3}{10} + \frac{2}{3} + \frac{2}{4} =$

27). $\frac{2}{3} + \frac{1}{10} + \frac{3}{4} =$

28). $\frac{2}{4} + \frac{1}{3} + \frac{2}{10} =$

29). $\frac{1}{3} + \frac{1}{2} + \frac{2}{10} =$

30). $\frac{1}{2} + \frac{4}{10} + \frac{3}{5} =$

31). $\frac{1}{4} + \frac{1}{2} + \frac{3}{10} =$

32). $\frac{2}{4} + \frac{9}{10} + \frac{1}{2} =$

33). $\frac{3}{4} + \frac{1}{3} + \frac{2}{5} =$

34). $\frac{3}{5} + \frac{9}{10} + \frac{1}{4} =$

35). $\frac{9}{10} + \frac{2}{4} + \frac{2}{3} =$

36). $\frac{1}{5} + \frac{1}{2} + \frac{2}{3} =$

37). $\frac{1}{2} + \frac{5}{10} + \frac{1}{3} =$

38). $\frac{7}{10} + \frac{1}{4} + \frac{1}{2} =$

39). $\frac{6}{10} + \frac{1}{2} + \frac{2}{4} =$

40). $\frac{1}{2} + \frac{4}{5} + \frac{2}{3} =$

41). $\frac{1}{3} + \frac{5}{8} + \frac{3}{4} =$

42). $\frac{7}{13} + \frac{12}{26} + \frac{5}{26} =$

43). $\frac{5}{6} + \frac{9}{18} + \frac{2}{18} =$

44). $\frac{6}{7} + \frac{2}{4} + \frac{10}{28} =$

45). $\frac{9}{22} + \frac{1}{11} + \frac{2}{11} =$

46). $\frac{3}{14} + \frac{3}{7} + \frac{3}{7} =$

47). $\frac{3}{5} + \frac{2}{10} + \frac{4}{5} =$

48). $\frac{1}{4} + \frac{9}{10} + \frac{3}{10} =$

49). $\frac{1}{11} + \frac{12}{22} + \frac{2}{22} =$

50). $\frac{3}{16} + \frac{7}{8} + \frac{11}{16} =$

51). $\frac{6}{18} + \frac{2}{6} + \frac{4}{6} =$

52). $\frac{1}{4} + \frac{1}{8} + \frac{11}{16} =$

53). $\frac{10}{26} + \frac{5}{13} + \frac{10}{13} =$

54). $\frac{1}{9} + \frac{7}{27} + \frac{1}{9} =$

55). $\frac{1}{7} + \frac{12}{21} + \frac{9}{21} =$

56). $\frac{10}{21} + \frac{4}{7} + \frac{5}{7} =$

57). $\frac{5}{20} + \frac{9}{10} + \frac{1}{4} =$

58). $\frac{2}{12} + \frac{4}{6} + \frac{2}{4} =$

59). $\frac{5}{28} + \frac{6}{7} + \frac{5}{7} =$

60). $\frac{4}{14} + \frac{1}{7} + \frac{3}{7} =$

61). $\frac{14}{46} + \frac{2}{23} + \frac{8}{46} =$

62). $\frac{2}{22} + \frac{3}{11} + \frac{4}{44} =$

63). $\frac{2}{6} + \frac{6}{18} + \frac{3}{12} =$

64). $\frac{5}{9} + \frac{12}{18} + \frac{7}{9} =$

65). $\frac{2}{7} + \frac{12}{14} + \frac{2}{7} =$

66). $\frac{7}{8} + \frac{3}{4} + \frac{1}{8} =$

67). $\frac{3}{4} + \frac{7}{12} + \frac{5}{6} =$

68). $\frac{4}{14} + \frac{2}{6} + \frac{6}{42} =$

69). $\frac{2}{3} + \frac{2}{4} + \frac{2}{3} =$

70). $\frac{13}{58} + \frac{2}{29} + \frac{6}{29} =$

71). $\frac{16}{22} + \frac{13}{44} + \frac{3}{11} =$

72). $\frac{13}{45} + \frac{3}{5} + \frac{3}{5} =$

73). $\frac{1}{5} + \frac{15}{20} + \frac{4}{20} =$

74). $\frac{2}{6} + \frac{1}{4} + \frac{1}{9} =$

75). $\frac{8}{24} + \frac{1}{48} + \frac{11}{16} =$

76). $\frac{4}{8} + \frac{6}{14} + \frac{11}{14} =$

77). $\frac{8}{10} + \frac{4}{5} + \frac{7}{10} =$

78). $\frac{1}{8} + \frac{3}{4} + \frac{9}{14} =$

79). $\frac{3}{9} + \frac{1}{3} + \frac{2}{6} =$

80). $\frac{5}{32} + \frac{7}{16} + \frac{7}{8} =$

81). $\frac{4}{16} + \frac{1}{4} + \frac{3}{4} =$

82). $\frac{4}{62} + \frac{7}{31} + \frac{20}{62} =$

83). $\frac{12}{39} + \frac{11}{78} + \frac{9}{13} =$

84). $\frac{4}{6} + \frac{12}{13} + \frac{3}{6} =$

85). $\frac{13}{72} + \frac{1}{3} + \frac{3}{72} =$

86). $\frac{8}{23} + \frac{10}{46} + \frac{8}{23} =$

87). $\frac{4}{46} + \frac{6}{23} + \frac{18}{46} =$

88). $\frac{9}{10} + \frac{1}{25} + \frac{8}{25} =$

89). $\frac{13}{29} + \frac{7}{58} + \frac{2}{58} =$

90). $\frac{6}{31} + \frac{9}{62} + \frac{4}{62} =$

91). $\frac{4}{5} + \frac{2}{14} + \frac{12}{14} =$

92). $\frac{7}{15} + \frac{1}{45} + \frac{1}{5} =$

93). $\frac{19}{37} + \frac{13}{74} + \frac{1}{74} =$

94). $\frac{6}{16} + \frac{3}{4} + \frac{13}{32} =$

95). $\frac{6}{25} + \frac{7}{10} + \frac{1}{4} =$

96). $\frac{3}{20} + \frac{3}{4} + \frac{4}{20} =$

97). $\frac{1}{22} + \frac{5}{11} + \frac{9}{22} =$

98). $\frac{1}{3} + \frac{12}{15} + \frac{2}{30} =$

99). $\frac{2}{3} + \frac{2}{5} + \frac{2}{3} =$

100). $\frac{2}{32} + \frac{2}{4} + \frac{7}{32} =$

ANSWER

1). $\dfrac{5}{7} + \dfrac{8}{7} + \dfrac{5}{7} = \dfrac{18}{7}$

2). $\dfrac{7}{7} + \dfrac{4}{7} + \dfrac{4}{7} = \dfrac{15}{7}$

3). $\dfrac{3}{5} + \dfrac{7}{5} + \dfrac{4}{5} = \dfrac{14}{5}$

4). $\dfrac{1}{12} + \dfrac{4}{12} + \dfrac{1}{12} = \dfrac{6}{12}$

5). $\dfrac{1}{6} + \dfrac{2}{6} + \dfrac{2}{6} = \dfrac{5}{6}$

6). $\dfrac{3}{7} + \dfrac{3}{7} + \dfrac{1}{7} = \dfrac{7}{7}$

7). $\dfrac{2}{7} + \dfrac{6}{7} + \dfrac{2}{7} = \dfrac{10}{7}$

8). $\dfrac{7}{9} + \dfrac{4}{9} + \dfrac{6}{9} = \dfrac{17}{9}$

9). $\dfrac{1}{7} + \dfrac{7}{7} + \dfrac{2}{7} = \dfrac{10}{7}$

10). $\dfrac{9}{8} + \dfrac{3}{8} + \dfrac{2}{8} = \dfrac{14}{8}$

11). $\dfrac{5}{8} + \dfrac{4}{8} + \dfrac{7}{8} = \dfrac{16}{8}$

12). $\dfrac{1}{9} + \dfrac{3}{9} + \dfrac{7}{9} = \dfrac{11}{9}$

13). $\dfrac{2}{8} + \dfrac{4}{8} + \dfrac{6}{8} = \dfrac{12}{8}$

14). $\dfrac{6}{7} + \dfrac{6}{7} + \dfrac{6}{7} = \dfrac{18}{7}$

15). $\dfrac{4}{11} + \dfrac{2}{11} + \dfrac{6}{11} = \dfrac{12}{11}$

16). $\dfrac{2}{9} + \dfrac{8}{9} + \dfrac{7}{9} = \dfrac{17}{9}$

17). $\dfrac{5}{8} + \dfrac{3}{8} + \dfrac{6}{8} = \dfrac{14}{8}$

18). $\dfrac{5}{4} + \dfrac{4}{4} + \dfrac{3}{4} = \dfrac{12}{4}$

19). $\dfrac{7}{11} + \dfrac{3}{11} + \dfrac{1}{11} = \dfrac{11}{11}$

20). $\dfrac{6}{6} + \dfrac{3}{6} + \dfrac{1}{6} = \dfrac{10}{6}$

21). $\dfrac{1}{4} + \dfrac{9}{10} + \dfrac{1}{5} = \dfrac{27}{20}$

22). $\dfrac{2}{5} + \dfrac{1}{3} + \dfrac{1}{2} = \dfrac{37}{30}$

23). $\dfrac{1}{3} + \dfrac{1}{2} + \dfrac{2}{5} = \dfrac{37}{30}$

24). $\dfrac{2}{4} + \dfrac{2}{3} + \dfrac{1}{2} = \dfrac{20}{12}$

25). $\dfrac{1}{2} + \dfrac{3}{5} + \dfrac{2}{3} = \dfrac{53}{30}$

26). $\dfrac{3}{10} + \dfrac{2}{3} + \dfrac{2}{4} = \dfrac{88}{60}$

27). $\dfrac{2}{3} + \dfrac{1}{10} + \dfrac{3}{4} = \dfrac{91}{60}$

28). $\dfrac{2}{4} + \dfrac{1}{3} + \dfrac{2}{10} = \dfrac{62}{60}$

29). $\dfrac{1}{3} + \dfrac{1}{2} + \dfrac{2}{10} = \dfrac{31}{30}$

30). $\dfrac{1}{2} + \dfrac{4}{10} + \dfrac{3}{5} = \dfrac{15}{10}$

31). $\dfrac{1}{4} + \dfrac{1}{2} + \dfrac{3}{10} = \dfrac{21}{20}$

32). $\dfrac{2}{4} + \dfrac{9}{10} + \dfrac{1}{2} = \dfrac{38}{20}$

33). $\dfrac{3}{4} + \dfrac{1}{3} + \dfrac{2}{5} = \dfrac{89}{60}$

34). $\dfrac{3}{5} + \dfrac{9}{10} + \dfrac{1}{4} = \dfrac{35}{20}$

35). $\dfrac{9}{10} + \dfrac{2}{4} + \dfrac{2}{3} = \dfrac{124}{60}$

36). $\dfrac{1}{5} + \dfrac{1}{2} + \dfrac{2}{3} = \dfrac{41}{30}$

37). $\dfrac{1}{2} + \dfrac{5}{10} + \dfrac{1}{3} = \dfrac{40}{30}$

38). $\dfrac{7}{10} + \dfrac{1}{4} + \dfrac{1}{2} = \dfrac{29}{20}$

39). $\dfrac{6}{10} + \dfrac{1}{2} + \dfrac{2}{4} = \dfrac{32}{20}$

40). $\dfrac{1}{2} + \dfrac{4}{5} + \dfrac{2}{3} = \dfrac{59}{30}$

41). $\dfrac{1}{3} + \dfrac{5}{8} + \dfrac{3}{4} = \dfrac{41}{24}$

51). $\dfrac{6}{18} + \dfrac{2}{6} + \dfrac{4}{6} = \dfrac{24}{18}$

42). $\dfrac{7}{13} + \dfrac{12}{26} + \dfrac{5}{26} = \dfrac{31}{26}$

52). $\dfrac{1}{4} + \dfrac{1}{8} + \dfrac{11}{16} = \dfrac{17}{16}$

43). $\dfrac{5}{6} + \dfrac{9}{18} + \dfrac{2}{18} = \dfrac{26}{18}$

53). $\dfrac{10}{26} + \dfrac{5}{13} + \dfrac{10}{13} = \dfrac{40}{26}$

44). $\dfrac{6}{7} + \dfrac{2}{4} + \dfrac{10}{28} = \dfrac{48}{28}$

54). $\dfrac{1}{9} + \dfrac{7}{27} + \dfrac{1}{9} = \dfrac{13}{27}$

45). $\dfrac{9}{22} + \dfrac{1}{11} + \dfrac{2}{11} = \dfrac{15}{22}$

55). $\dfrac{1}{7} + \dfrac{12}{21} + \dfrac{9}{21} = \dfrac{24}{21}$

46). $\dfrac{3}{14} + \dfrac{3}{7} + \dfrac{3}{7} = \dfrac{15}{14}$

56). $\dfrac{10}{21} + \dfrac{4}{7} + \dfrac{5}{7} = \dfrac{37}{21}$

47). $\dfrac{3}{5} + \dfrac{2}{10} + \dfrac{4}{5} = \dfrac{16}{10}$

57). $\dfrac{5}{20} + \dfrac{9}{10} + \dfrac{1}{4} = \dfrac{28}{20}$

48). $\dfrac{1}{4} + \dfrac{9}{10} + \dfrac{3}{10} = \dfrac{29}{20}$

58). $\dfrac{2}{12} + \dfrac{4}{6} + \dfrac{2}{4} = \dfrac{16}{12}$

49). $\dfrac{1}{11} + \dfrac{12}{22} + \dfrac{2}{22} = \dfrac{16}{22}$

59). $\dfrac{5}{28} + \dfrac{6}{7} + \dfrac{5}{7} = \dfrac{49}{28}$

50). $\dfrac{3}{16} + \dfrac{7}{8} + \dfrac{11}{16} = \dfrac{28}{16}$

60). $\dfrac{4}{14} + \dfrac{1}{7} + \dfrac{3}{7} = \dfrac{12}{14}$

61). $\frac{14}{46} + \frac{2}{23} + \frac{8}{46} = \frac{26}{46}$

62). $\frac{2}{22} + \frac{3}{11} + \frac{4}{44} = \frac{20}{44}$

63). $\frac{2}{6} + \frac{6}{18} + \frac{3}{12} = \frac{33}{36}$

64). $\frac{5}{9} + \frac{12}{18} + \frac{7}{9} = \frac{36}{18}$

65). $\frac{2}{7} + \frac{12}{14} + \frac{2}{7} = \frac{20}{14}$

66). $\frac{7}{8} + \frac{3}{4} + \frac{1}{8} = \frac{14}{8}$

67). $\frac{3}{4} + \frac{7}{12} + \frac{5}{6} = \frac{26}{12}$

68). $\frac{4}{14} + \frac{2}{6} + \frac{6}{42} = \frac{32}{42}$

69). $\frac{2}{3} + \frac{2}{4} + \frac{2}{3} = \frac{22}{12}$

70). $\frac{13}{58} + \frac{2}{29} + \frac{6}{29} = \frac{29}{58}$

71). $\frac{16}{22} + \frac{13}{44} + \frac{3}{11} = \frac{57}{44}$

72). $\frac{13}{45} + \frac{3}{5} + \frac{3}{5} = \frac{67}{45}$

73). $\frac{1}{5} + \frac{15}{20} + \frac{4}{20} = \frac{23}{20}$

74). $\frac{2}{6} + \frac{1}{4} + \frac{1}{9} = \frac{25}{36}$

75). $\frac{8}{24} + \frac{1}{48} + \frac{11}{16} = \frac{50}{48}$

76). $\frac{4}{8} + \frac{6}{14} + \frac{11}{14} = \frac{96}{56}$

77). $\frac{8}{10} + \frac{4}{5} + \frac{7}{10} = \frac{23}{10}$

78). $\frac{1}{8} + \frac{3}{4} + \frac{9}{14} = \frac{85}{56}$

79). $\frac{3}{9} + \frac{1}{3} + \frac{2}{6} = \frac{18}{18}$

80). $\frac{5}{32} + \frac{7}{16} + \frac{7}{8} = \frac{47}{32}$

81). $\dfrac{4}{16} + \dfrac{1}{4} + \dfrac{3}{4} = \dfrac{20}{16}$

91). $\dfrac{4}{5} + \dfrac{2}{14} + \dfrac{12}{14} = \dfrac{126}{70}$

82). $\dfrac{4}{62} + \dfrac{7}{31} + \dfrac{20}{62} = \dfrac{38}{62}$

92). $\dfrac{7}{15} + \dfrac{1}{45} + \dfrac{1}{5} = \dfrac{31}{45}$

83). $\dfrac{12}{39} + \dfrac{11}{78} + \dfrac{9}{13} = \dfrac{89}{78}$

93). $\dfrac{19}{37} + \dfrac{13}{74} + \dfrac{1}{74} = \dfrac{52}{74}$

84). $\dfrac{4}{6} + \dfrac{12}{13} + \dfrac{3}{6} = \dfrac{163}{78}$

94). $\dfrac{6}{16} + \dfrac{3}{4} + \dfrac{13}{32} = \dfrac{49}{32}$

85). $\dfrac{13}{72} + \dfrac{1}{3} + \dfrac{3}{72} = \dfrac{40}{72}$

95). $\dfrac{6}{25} + \dfrac{7}{10} + \dfrac{1}{4} = \dfrac{119}{100}$

86). $\dfrac{8}{23} + \dfrac{10}{46} + \dfrac{8}{23} = \dfrac{42}{46}$

96). $\dfrac{3}{20} + \dfrac{3}{4} + \dfrac{4}{20} = \dfrac{22}{20}$

87). $\dfrac{4}{46} + \dfrac{6}{23} + \dfrac{18}{46} = \dfrac{34}{46}$

97). $\dfrac{1}{22} + \dfrac{5}{11} + \dfrac{9}{22} = \dfrac{20}{22}$

88). $\dfrac{9}{10} + \dfrac{1}{25} + \dfrac{8}{25} = \dfrac{63}{50}$

98). $\dfrac{1}{3} + \dfrac{12}{15} + \dfrac{2}{30} = \dfrac{36}{30}$

89). $\dfrac{13}{29} + \dfrac{7}{58} + \dfrac{2}{58} = \dfrac{35}{58}$

99). $\dfrac{2}{3} + \dfrac{2}{5} + \dfrac{2}{3} = \dfrac{26}{15}$

90). $\dfrac{6}{31} + \dfrac{9}{62} + \dfrac{4}{62} = \dfrac{25}{62}$

100). $\dfrac{2}{32} + \dfrac{2}{4} + \dfrac{7}{32} = \dfrac{25}{32}$

ADDING MIXED NUMBERS FRACTIONS

1). $4\frac{1}{4} + 8\frac{1}{3} =$

2). $3\frac{1}{3} + 9\frac{1}{2} =$

3). $2\frac{7}{10} + 9\frac{1}{3} =$

4). $6\frac{1}{2} + 5\frac{5}{10} =$

5). $1\frac{2}{3} + 4\frac{1}{4} =$

6). $6\frac{2}{4} + 9\frac{1}{2} =$

7). $5\frac{1}{2} + 9\frac{6}{10} =$

8). $5\frac{1}{3} + 6\frac{1}{2} =$

9). $2\frac{4}{5} + 9\frac{2}{3} =$

10). $1\frac{2}{3} + 4\frac{3}{4} =$

11). $5\frac{4}{5} + 7\frac{1}{4} =$

12). $1\frac{1}{2} + 9\frac{2}{5} =$

13). $2\frac{1}{2} + 4\frac{2}{3} =$

14). $3\frac{1}{2} + 4\frac{3}{10} =$

15). $2\frac{1}{2} + 6\frac{2}{5} =$

16). $6\frac{1}{2} + 9\frac{3}{5} =$

17). $1\frac{2}{4} + 7\frac{1}{10} =$

18). $2\frac{2}{3} + 7\frac{1}{2} =$

19). $2\frac{2}{3} + 7\frac{3}{5} =$

20). $2\frac{2}{4} + 7\frac{1}{2} =$

21). $3\frac{5}{10} + 5\frac{2}{4} =$

22). $2\frac{1}{5} + 7\frac{1}{2} =$

23). $2\frac{4}{5} + 8\frac{3}{4} =$

24). $5\frac{2}{10} + 6\frac{1}{5} =$

25). $1\frac{1}{2} + 5\frac{1}{4} =$

26). $2\frac{2}{3} + 8\frac{3}{10} =$

27). $2\frac{1}{4} + 4\frac{1}{3} =$

28). $1\frac{1}{2} + 6\frac{1}{10} =$

29). $6\frac{2}{4} + 4\frac{2}{3} =$

30). $3\frac{9}{10} + 8\frac{1}{2} =$

31). $4\frac{3}{4} + 9\frac{2}{5} =$

32). $5\frac{1}{3} + 9\frac{1}{4} =$

33). $4\frac{1}{3} + 9\frac{1}{2} =$

34). $6\frac{4}{10} + 6\frac{2}{4} =$

35). $2\frac{4}{5} + 4\frac{2}{3} =$

36). $1\frac{1}{2} + 8\frac{6}{10} =$

37). $2\frac{2}{3} + 6\frac{2}{5} =$

38). $6\frac{2}{10} + 9\frac{1}{2} =$

39). $2\frac{1}{2} + 5\frac{1}{3} =$

40). $5\frac{4}{5} + 9\frac{2}{3} =$

41). $5\frac{3}{4} + 7\frac{2}{3} =$

42). $1\frac{1}{2} + 7\frac{1}{3} =$

43). $3\frac{1}{2} + 6\frac{2}{4} =$

44). $3\frac{2}{10} + 7\frac{1}{3} =$

45). $4\frac{1}{2} + 8\frac{2}{4} =$

46). $1\frac{1}{2} + 8\frac{3}{10} =$

47). $5\frac{1}{4} + 4\frac{1}{10} =$

48). $4\frac{1}{2} + 9\frac{1}{3} =$

49). $1\frac{1}{2} + 4\frac{2}{4} =$

50). $2\frac{2}{3} + 8\frac{1}{2} =$

ANSWER

1). $4\frac{1}{4} + 8\frac{1}{3} = 12\frac{7}{12}$

11). $5\frac{4}{5} + 7\frac{1}{4} = 12\frac{21}{20}$

2). $3\frac{1}{3} + 9\frac{1}{2} = 12\frac{5}{6}$

12). $1\frac{1}{2} + 9\frac{2}{5} = 10\frac{9}{10}$

3). $2\frac{7}{10} + 9\frac{1}{3} = 11\frac{31}{30}$

13). $2\frac{1}{2} + 4\frac{2}{3} = 6\frac{7}{6}$

4). $6\frac{1}{2} + 5\frac{5}{10} = 11\frac{10}{10}$

14). $3\frac{1}{2} + 4\frac{3}{10} = 7\frac{8}{10}$

5). $1\frac{2}{3} + 4\frac{1}{4} = 5\frac{11}{12}$

15). $2\frac{1}{2} + 6\frac{2}{5} = 8\frac{9}{10}$

6). $6\frac{2}{4} + 9\frac{1}{2} = 15\frac{4}{4}$

16). $6\frac{1}{2} + 9\frac{3}{5} = 15\frac{11}{10}$

7). $5\frac{1}{2} + 9\frac{6}{10} = 14\frac{11}{10}$

17). $1\frac{2}{4} + 7\frac{1}{10} = 8\frac{12}{20}$

8). $5\frac{1}{3} + 6\frac{1}{2} = 11\frac{5}{6}$

18). $2\frac{2}{3} + 7\frac{1}{2} = 9\frac{7}{6}$

9). $2\frac{4}{5} + 9\frac{2}{3} = 11\frac{22}{15}$

19). $2\frac{2}{3} + 7\frac{3}{5} = 9\frac{19}{15}$

10). $1\frac{2}{3} + 4\frac{3}{4} = 5\frac{17}{12}$

20). $2\frac{2}{4} + 7\frac{1}{2} = 9\frac{4}{4}$

21). $3\frac{5}{10} + 5\frac{2}{4} = 8\frac{20}{20}$

22). $2\frac{1}{5} + 7\frac{1}{2} = 9\frac{7}{10}$

23). $2\frac{4}{5} + 8\frac{3}{4} = 10\frac{31}{20}$

24). $5\frac{2}{10} + 6\frac{1}{5} = 11\frac{4}{10}$

25). $1\frac{1}{2} + 5\frac{1}{4} = 6\frac{3}{4}$

26). $2\frac{2}{3} + 8\frac{3}{10} = 10\frac{29}{30}$

27). $2\frac{1}{4} + 4\frac{1}{3} = 6\frac{7}{12}$

28). $1\frac{1}{2} + 6\frac{1}{10} = 7\frac{6}{10}$

29). $6\frac{2}{4} + 4\frac{2}{3} = 10\frac{14}{12}$

30). $3\frac{9}{10} + 8\frac{1}{2} = 11\frac{14}{10}$

31). $4\frac{3}{4} + 9\frac{2}{5} = 13\frac{23}{20}$

32). $5\frac{1}{3} + 9\frac{1}{4} = 14\frac{7}{12}$

33). $4\frac{1}{3} + 9\frac{1}{2} = 13\frac{5}{6}$

34). $6\frac{4}{10} + 6\frac{2}{4} = 12\frac{18}{20}$

35). $2\frac{4}{5} + 4\frac{2}{3} = 6\frac{22}{15}$

36). $1\frac{1}{2} + 8\frac{6}{10} = 9\frac{11}{10}$

37). $2\frac{2}{3} + 6\frac{2}{5} = 8\frac{16}{15}$

38). $6\frac{2}{10} + 9\frac{1}{2} = 15\frac{7}{10}$

39). $2\frac{1}{2} + 5\frac{1}{3} = 7\frac{5}{6}$

40). $5\frac{4}{5} + 9\frac{2}{3} = 14\frac{22}{15}$

41). $5\frac{3}{4} + 7\frac{2}{3} = 12\frac{17}{12}$

42). $1\frac{1}{2} + 7\frac{1}{3} = 8\frac{5}{6}$

43). $3\frac{1}{2} + 6\frac{2}{4} = 9\frac{4}{4}$

44). $3\frac{2}{10} + 7\frac{1}{3} = 10\frac{16}{30}$

45). $4\frac{1}{2} + 8\frac{2}{4} = 12\frac{4}{4}$

46). $1\frac{1}{2} + 8\frac{3}{10} = 9\frac{8}{10}$

47). $5\frac{1}{4} + 4\frac{1}{10} = 9\frac{7}{20}$

48). $4\frac{1}{2} + 9\frac{1}{3} = 13\frac{5}{6}$

49). $1\frac{1}{2} + 4\frac{2}{4} = 5\frac{4}{4}$

50). $2\frac{2}{3} + 8\frac{1}{2} = 10\frac{7}{6}$

SUBTRACTION OF FRACTIONS

1). $\frac{10}{12} - \frac{2}{12} =$

2). $\frac{9}{11} - \frac{5}{11} =$

3). $\frac{2}{4} - \frac{1}{4} =$

4). $\frac{2}{8} - \frac{1}{8} =$

5). $\frac{2}{3} - \frac{1}{3} =$

6). $\frac{4}{5} - \frac{2}{5} =$

7). $\frac{5}{10} - \frac{2}{10} =$

8). $\frac{3}{10} - \frac{2}{10} =$

9). $\frac{3}{6} - \frac{2}{6} =$

10). $\frac{8}{12} - \frac{4}{12} =$

11). $\frac{7}{10} - \frac{1}{10} =$

12). $\frac{2}{9} - \frac{1}{9} =$

13). $\frac{2}{8} - \frac{1}{8} =$

14). $\frac{2}{5} - \frac{1}{5} =$

15). $\frac{5}{6} - \frac{1}{6} =$

16). $\frac{3}{9} - \frac{2}{9} =$

17). $\frac{6}{11} - \frac{4}{11} =$

18). $\frac{3}{4} - \frac{2}{4} =$

19). $\frac{2}{10} - \frac{1}{10} =$

20). $\frac{8}{11} - \frac{5}{11} =$

21). $\dfrac{4}{12} - \dfrac{3}{12} =$

22). $\dfrac{2}{10} - \dfrac{1}{10} =$

23). $\dfrac{2}{4} - \dfrac{1}{4} =$

24). $\dfrac{3}{6} - \dfrac{2}{6} =$

25). $\dfrac{3}{11} - \dfrac{2}{11} =$

26). $\dfrac{8}{12} - \dfrac{7}{12} =$

27). $\dfrac{2}{9} - \dfrac{1}{9} =$

28). $\dfrac{9}{12} - \dfrac{6}{12} =$

29). $\dfrac{5}{8} - \dfrac{2}{8} =$

30). $\dfrac{6}{11} - \dfrac{2}{11} =$

31). $\dfrac{3}{5} - \dfrac{1}{5} =$

32). $\dfrac{2}{3} - \dfrac{1}{3} =$

33). $\dfrac{4}{7} - \dfrac{3}{7} =$

34). $\dfrac{3}{6} - \dfrac{2}{6} =$

35). $\dfrac{7}{9} - \dfrac{1}{9} =$

36). $\dfrac{3}{11} - \dfrac{2}{11} =$

37). $\dfrac{5}{10} - \dfrac{4}{10} =$

38). $\dfrac{3}{4} - \dfrac{2}{4} =$

39). $\dfrac{11}{12} - \dfrac{10}{12} =$

40). $\dfrac{4}{8} - \dfrac{2}{8} =$

41). $\dfrac{6}{9} - \dfrac{5}{9} =$

42). $\dfrac{6}{7} - \dfrac{3}{7} =$

43). $\dfrac{4}{12} - \dfrac{3}{12} =$

44). $\dfrac{2}{10} - \dfrac{1}{10} =$

45). $\dfrac{4}{8} - \dfrac{3}{8} =$

46). $\dfrac{3}{6} - \dfrac{2}{6} =$

47). $\dfrac{3}{5} - \dfrac{1}{5} =$

48). $\dfrac{8}{11} - \dfrac{1}{11} =$

49). $\dfrac{2}{3} - \dfrac{1}{3} =$

50). $\dfrac{5}{9} - \dfrac{2}{9} =$

51). $\dfrac{10}{12} - \dfrac{2}{12} =$

52). $\dfrac{8}{11} - \dfrac{2}{11} =$

53). $\dfrac{11}{12} - \dfrac{6}{12} =$

54). $\dfrac{3}{6} - \dfrac{2}{6} =$

55). $\dfrac{2}{4} - \dfrac{1}{4} =$

56). $\dfrac{3}{7} - \dfrac{2}{7} =$

57). $\dfrac{6}{9} - \dfrac{2}{9} =$

58). $\dfrac{2}{12} - \dfrac{1}{12} =$

59). $\dfrac{6}{9} - \dfrac{1}{9} =$

60). $\dfrac{9}{10} - \dfrac{8}{10} =$

61). $\dfrac{2}{3} - \dfrac{1}{4} =$

62). $\dfrac{1}{2} - \dfrac{1}{10} =$

63). $\dfrac{1}{2} - \dfrac{2}{4} =$

64). $\dfrac{5}{10} - \dfrac{1}{5} =$

65). $\dfrac{1}{3} - \dfrac{1}{5} =$

66). $\dfrac{4}{10} - \dfrac{1}{3} =$

67). $\dfrac{1}{3} - \dfrac{3}{10} =$

68). $\dfrac{3}{5} - \dfrac{1}{3} =$

69). $\dfrac{2}{5} - \dfrac{1}{4} =$

70). $\dfrac{9}{10} - \dfrac{1}{3} =$

71). $\dfrac{4}{5} - \dfrac{1}{4} =$

72). $\dfrac{7}{10} - \dfrac{3}{5} =$

73). $\dfrac{1}{2} - \dfrac{2}{10} =$

74). $\dfrac{1}{2} - \dfrac{3}{10} =$

75). $\dfrac{8}{10} - \dfrac{3}{4} =$

76). $\dfrac{3}{5} - \dfrac{2}{10} =$

77). $\dfrac{3}{5} - \dfrac{1}{2} =$

78). $\dfrac{2}{3} - \dfrac{3}{10} =$

79). $\dfrac{3}{4} - \dfrac{1}{2} =$

80). $\dfrac{2}{3} - \dfrac{1}{4} =$

81). $\frac{1}{3} - \frac{2}{6} =$

82). $\frac{6}{7} - \frac{4}{21} =$

83). $\frac{7}{8} - \frac{5}{16} =$

84). $\frac{7}{11} - \frac{11}{22} =$

85). $\frac{9}{13} - \frac{4}{26} =$

86). $\frac{10}{14} - \frac{4}{7} =$

87). $\frac{9}{11} - \frac{8}{22} =$

88). $\frac{7}{11} - \frac{7}{22} =$

89). $\frac{11}{16} - \frac{2}{4} =$

90). $\frac{3}{6} - \frac{3}{12} =$

91). $\frac{8}{12} - \frac{1}{4} =$

92). $\frac{4}{7} - \frac{11}{21} =$

93). $\frac{5}{13} - \frac{10}{26} =$

94). $\frac{2}{5} - \frac{1}{4} =$

95). $\frac{2}{3} - \frac{8}{18} =$

96). $\frac{1}{3} - \frac{2}{24} =$

97). $\frac{6}{12} - \frac{1}{3} =$

98). $\frac{8}{9} - \frac{2}{27} =$

99). $\frac{6}{7} - \frac{3}{14} =$

100). $\frac{10}{12} - \frac{2}{3} =$

ANSWER

1). $\dfrac{10}{12} - \dfrac{2}{12} = \dfrac{8}{12}$

2). $\dfrac{9}{11} - \dfrac{5}{11} = \dfrac{4}{11}$

3). $\dfrac{2}{4} - \dfrac{1}{4} = \dfrac{1}{4}$

4). $\dfrac{2}{8} - \dfrac{1}{8} = \dfrac{1}{8}$

5). $\dfrac{2}{3} - \dfrac{1}{3} = \dfrac{1}{3}$

6). $\dfrac{4}{5} - \dfrac{2}{5} = \dfrac{2}{5}$

7). $\dfrac{5}{10} - \dfrac{2}{10} = \dfrac{3}{10}$

8). $\dfrac{3}{10} - \dfrac{2}{10} = \dfrac{1}{10}$

9). $\dfrac{3}{6} - \dfrac{2}{6} = \dfrac{1}{6}$

10). $\dfrac{8}{12} - \dfrac{4}{12} = \dfrac{4}{12}$

11). $\dfrac{7}{10} - \dfrac{1}{10} = \dfrac{6}{10}$

12). $\dfrac{2}{9} - \dfrac{1}{9} = \dfrac{1}{9}$

13). $\dfrac{2}{8} - \dfrac{1}{8} = \dfrac{1}{8}$

14). $\dfrac{2}{5} - \dfrac{1}{5} = \dfrac{1}{5}$

15). $\dfrac{5}{6} - \dfrac{1}{6} = \dfrac{4}{6}$

16). $\dfrac{3}{9} - \dfrac{2}{9} = \dfrac{1}{9}$

17). $\dfrac{6}{11} - \dfrac{4}{11} = \dfrac{2}{11}$

18). $\dfrac{3}{4} - \dfrac{2}{4} = \dfrac{1}{4}$

19). $\dfrac{2}{10} - \dfrac{1}{10} = \dfrac{1}{10}$

20). $\dfrac{8}{11} - \dfrac{5}{11} = \dfrac{3}{11}$

21). $\frac{4}{12} - \frac{3}{12} = \frac{1}{12}$

22). $\frac{2}{10} - \frac{1}{10} = \frac{1}{10}$

23). $\frac{2}{4} - \frac{1}{4} = \frac{1}{4}$

24). $\frac{3}{6} - \frac{2}{6} = \frac{1}{6}$

25). $\frac{3}{11} - \frac{2}{11} = \frac{1}{11}$

26). $\frac{8}{12} - \frac{7}{12} = \frac{1}{12}$

27). $\frac{2}{9} - \frac{1}{9} = \frac{1}{9}$

28). $\frac{9}{12} - \frac{6}{12} = \frac{3}{12}$

29). $\frac{5}{8} - \frac{2}{8} = \frac{3}{8}$

30). $\frac{6}{11} - \frac{2}{11} = \frac{4}{11}$

31). $\frac{3}{5} - \frac{1}{5} = \frac{2}{5}$

32). $\frac{2}{3} - \frac{1}{3} = \frac{1}{3}$

33). $\frac{4}{7} - \frac{3}{7} = \frac{1}{7}$

34). $\frac{3}{6} - \frac{2}{6} = \frac{1}{6}$

35). $\frac{7}{9} - \frac{1}{9} = \frac{6}{9}$

36). $\frac{3}{11} - \frac{2}{11} = \frac{1}{11}$

37). $\frac{5}{10} - \frac{4}{10} = \frac{1}{10}$

38). $\frac{3}{4} - \frac{2}{4} = \frac{1}{4}$

39). $\frac{11}{12} - \frac{10}{12} = \frac{1}{12}$

40). $\frac{4}{8} - \frac{2}{8} = \frac{2}{8}$

41). $\dfrac{6}{9} - \dfrac{5}{9} = \dfrac{1}{9}$

42). $\dfrac{6}{7} - \dfrac{3}{7} = \dfrac{3}{7}$

43). $\dfrac{4}{12} - \dfrac{3}{12} = \dfrac{1}{12}$

44). $\dfrac{2}{10} - \dfrac{1}{10} = \dfrac{1}{10}$

45). $\dfrac{4}{8} - \dfrac{3}{8} = \dfrac{1}{8}$

46). $\dfrac{3}{6} - \dfrac{2}{6} = \dfrac{1}{6}$

47). $\dfrac{3}{5} - \dfrac{1}{5} = \dfrac{2}{5}$

48). $\dfrac{8}{11} - \dfrac{1}{11} = \dfrac{7}{11}$

49). $\dfrac{2}{3} - \dfrac{1}{3} = \dfrac{1}{3}$

50). $\dfrac{5}{9} - \dfrac{2}{9} = \dfrac{3}{9}$

51). $\dfrac{10}{12} - \dfrac{2}{12} = \dfrac{8}{12}$

52). $\dfrac{8}{11} - \dfrac{2}{11} = \dfrac{6}{11}$

53). $\dfrac{11}{12} - \dfrac{6}{12} = \dfrac{5}{12}$

54). $\dfrac{3}{6} - \dfrac{2}{6} = \dfrac{1}{6}$

55). $\dfrac{2}{4} - \dfrac{1}{4} = \dfrac{1}{4}$

56). $\dfrac{3}{7} - \dfrac{2}{7} = \dfrac{1}{7}$

57). $\dfrac{6}{9} - \dfrac{2}{9} = \dfrac{4}{9}$

58). $\dfrac{2}{12} - \dfrac{1}{12} = \dfrac{1}{12}$

59). $\dfrac{6}{9} - \dfrac{1}{9} = \dfrac{5}{9}$

60). $\dfrac{9}{10} - \dfrac{8}{10} = \dfrac{1}{10}$

61). $\dfrac{2}{3} - \dfrac{1}{4} = \dfrac{5}{12}$

62). $\dfrac{1}{2} - \dfrac{1}{10} = \dfrac{4}{10}$

63). $\dfrac{1}{2} - \dfrac{2}{4} = 0$

64). $\dfrac{5}{10} - \dfrac{1}{5} = \dfrac{3}{10}$

65). $\dfrac{1}{3} - \dfrac{1}{5} = \dfrac{2}{15}$

66). $\dfrac{4}{10} - \dfrac{1}{3} = \dfrac{2}{30}$

67). $\dfrac{1}{3} - \dfrac{3}{10} = \dfrac{1}{30}$

68). $\dfrac{3}{5} - \dfrac{1}{3} = \dfrac{4}{15}$

69). $\dfrac{2}{5} - \dfrac{1}{4} = \dfrac{3}{20}$

70). $\dfrac{9}{10} - \dfrac{1}{3} = \dfrac{17}{30}$

71). $\dfrac{4}{5} - \dfrac{1}{4} = \dfrac{11}{20}$

72). $\dfrac{7}{10} - \dfrac{3}{5} = \dfrac{1}{10}$

73). $\dfrac{1}{2} - \dfrac{2}{10} = \dfrac{3}{10}$

74). $\dfrac{1}{2} - \dfrac{3}{10} = \dfrac{2}{10}$

75). $\dfrac{8}{10} - \dfrac{3}{4} = \dfrac{1}{20}$

76). $\dfrac{3}{5} - \dfrac{2}{10} = \dfrac{4}{10}$

77). $\dfrac{3}{5} - \dfrac{1}{2} = \dfrac{1}{10}$

78). $\dfrac{2}{3} - \dfrac{3}{10} = \dfrac{11}{30}$

79). $\dfrac{3}{4} - \dfrac{1}{2} = \dfrac{1}{4}$

80). $\dfrac{2}{3} - \dfrac{1}{4} = \dfrac{5}{12}$

81). $\dfrac{1}{3} - \dfrac{2}{6} = 0$

82). $\dfrac{6}{7} - \dfrac{4}{21} = \dfrac{14}{21}$

83). $\dfrac{7}{8} - \dfrac{5}{16} = \dfrac{9}{16}$

84). $\dfrac{7}{11} - \dfrac{11}{22} = \dfrac{3}{22}$

85). $\dfrac{9}{13} - \dfrac{4}{26} = \dfrac{14}{26}$

86). $\dfrac{10}{14} - \dfrac{4}{7} = \dfrac{2}{14}$

87). $\dfrac{9}{11} - \dfrac{8}{22} = \dfrac{10}{22}$

88). $\dfrac{7}{11} - \dfrac{7}{22} = \dfrac{7}{22}$

89). $\dfrac{11}{16} - \dfrac{2}{4} = \dfrac{3}{16}$

90). $\dfrac{3}{6} - \dfrac{3}{12} = \dfrac{3}{12}$

91). $\dfrac{8}{12} - \dfrac{1}{4} = \dfrac{5}{12}$

92). $\dfrac{4}{7} - \dfrac{11}{21} = \dfrac{1}{21}$

93). $\dfrac{5}{13} - \dfrac{10}{26} = 0$

94). $\dfrac{2}{5} - \dfrac{1}{4} = \dfrac{3}{20}$

95). $\dfrac{2}{3} - \dfrac{8}{18} = \dfrac{4}{18}$

96). $\dfrac{1}{3} - \dfrac{2}{24} = \dfrac{6}{24}$

97). $\dfrac{6}{12} - \dfrac{1}{3} = \dfrac{2}{12}$

98). $\dfrac{8}{9} - \dfrac{2}{27} = \dfrac{22}{27}$

99). $\dfrac{6}{7} - \dfrac{3}{14} = \dfrac{9}{14}$

100). $\dfrac{10}{12} - \dfrac{2}{3} = \dfrac{2}{12}$

SUBTRACTION OF THREE FRACTIONS

1). $\dfrac{7}{8} - \dfrac{5}{8} - \dfrac{1}{8} =$

2). $\dfrac{7}{8} - \dfrac{2}{8} - \dfrac{1}{8} =$

3). $\dfrac{10}{11} - \dfrac{4}{11} - \dfrac{5}{11} =$

4). $\dfrac{10}{11} - \dfrac{4}{11} - \dfrac{1}{11} =$

5). $\dfrac{5}{6} - \dfrac{3}{6} - \dfrac{1}{6} =$

6). $\dfrac{10}{11} - \dfrac{7}{11} - \dfrac{2}{11} =$

7). $\dfrac{6}{7} - \dfrac{1}{7} - \dfrac{3}{7} =$

8). $\dfrac{5}{6} - \dfrac{2}{6} - \dfrac{2}{6} =$

9). $\dfrac{7}{8} - \dfrac{3}{8} - \dfrac{1}{8} =$

10). $\dfrac{5}{6} - \dfrac{3}{6} - \dfrac{1}{6} =$

11). $\dfrac{11}{12} - \dfrac{6}{12} - \dfrac{3}{12} =$

12). $\dfrac{9}{10} - \dfrac{1}{10} - \dfrac{6}{10} =$

13). $\dfrac{6}{7} - \dfrac{4}{7} - \dfrac{1}{7} =$

14). $\dfrac{8}{9} - \dfrac{5}{9} - \dfrac{1}{9} =$

15). $\dfrac{9}{10} - \dfrac{5}{10} - \dfrac{1}{10} =$

16). $\dfrac{8}{9} - \dfrac{1}{9} - \dfrac{4}{9} =$

17). $\dfrac{5}{6} - \dfrac{3}{6} - \dfrac{1}{6} =$

18). $\dfrac{9}{10} - \dfrac{5}{10} - \dfrac{3}{10} =$

19). $\dfrac{9}{10} - \dfrac{5}{10} - \dfrac{1}{10} =$

20). $\dfrac{7}{8} - \dfrac{5}{8} - \dfrac{1}{8} =$

21). $\frac{8}{10} - \frac{1}{2} - \frac{3}{10} =$

22). $\frac{4}{5} - \frac{1}{5} - \frac{1}{5} =$

23). $\frac{8}{10} - \frac{1}{3} - \frac{3}{10} =$

24). $\frac{4}{5} - \frac{2}{5} - \frac{3}{10} =$

25). $\frac{9}{10} - \frac{1}{4} - \frac{1}{5} =$

26). $\frac{4}{5} - \frac{1}{4} - \frac{1}{10} =$

27). $\frac{4}{5} - \frac{1}{2} - \frac{1}{5} =$

28). $\frac{8}{10} - \frac{1}{4} - \frac{3}{10} =$

29). $\frac{9}{10} - \frac{1}{5} - \frac{2}{10} =$

30). $\frac{9}{10} - \frac{1}{5} - \frac{2}{10} =$

31). $\frac{8}{10} - \frac{1}{5} - \frac{1}{10} =$

32). $\frac{9}{10} - \frac{1}{2} - \frac{1}{5} =$

33). $\frac{9}{10} - \frac{1}{5} - \frac{1}{5} =$

34). $\frac{4}{5} - \frac{1}{4} - \frac{1}{10} =$

35). $\frac{4}{5} - \frac{1}{3} - \frac{1}{10} =$

36). $\frac{4}{5} - \frac{1}{3} - \frac{1}{10} =$

37). $\frac{4}{5} - \frac{1}{2} - \frac{1}{5} =$

38). $\frac{4}{5} - \frac{1}{2} - \frac{1}{5} =$

39). $\frac{4}{5} - \frac{1}{3} - \frac{3}{10} =$

40). $\frac{9}{10} - \frac{1}{2} - \frac{3}{10} =$

41). $\dfrac{11}{12} - \dfrac{1}{3} - \dfrac{1}{6} =$

42). $\dfrac{11}{12} - \dfrac{1}{3} - \dfrac{1}{6} =$

43). $\dfrac{11}{12} - \dfrac{2}{4} - \dfrac{1}{6} =$

44). $\dfrac{23}{24} - \dfrac{2}{4} - \dfrac{1}{12} =$

45). $\dfrac{26}{28} - \dfrac{4}{7} - \dfrac{1}{14} =$

46). $\dfrac{18}{20} - \dfrac{3}{5} - \dfrac{1}{10} =$

47). $\dfrac{15}{16} - \dfrac{1}{4} - \dfrac{2}{8} =$

48). $\dfrac{22}{24} - \dfrac{2}{8} - \dfrac{2}{12} =$

49). $\dfrac{26}{28} - \dfrac{1}{4} - \dfrac{1}{14} =$

50). $\dfrac{11}{12} - \dfrac{1}{3} - \dfrac{1}{6} =$

51). $\dfrac{26}{28} - \dfrac{2}{7} - \dfrac{1}{14} =$

52). $\dfrac{11}{12} - \dfrac{1}{3} - \dfrac{1}{6} =$

53). $\dfrac{16}{18} - \dfrac{1}{3} - \dfrac{1}{9} =$

54). $\dfrac{22}{24} - \dfrac{4}{8} - \dfrac{2}{12} =$

55). $\dfrac{26}{28} - \dfrac{1}{4} - \dfrac{2}{14} =$

56). $\dfrac{28}{30} - \dfrac{4}{6} - \dfrac{2}{15} =$

57). $\dfrac{18}{20} - \dfrac{1}{4} - \dfrac{2}{10} =$

58). $\dfrac{29}{30} - \dfrac{1}{6} - \dfrac{1}{15} =$

59). $\dfrac{26}{28} - \dfrac{2}{7} - \dfrac{2}{14} =$

60). $\dfrac{29}{30} - \dfrac{4}{6} - \dfrac{1}{15} =$

61). $\dfrac{40}{42} - \dfrac{3}{6} - \dfrac{2}{21} =$

62). $\dfrac{49}{50} - \dfrac{5}{10} - \dfrac{1}{25} =$

63). $\dfrac{44}{45} - \dfrac{3}{5} - \dfrac{2}{15} =$

64). $\dfrac{59}{60} - \dfrac{1}{6} - \dfrac{1}{30} =$

65). $\dfrac{49}{50} - \dfrac{1}{5} - \dfrac{1}{25} =$

66). $\dfrac{39}{40} - \dfrac{2}{4} - \dfrac{1}{20} =$

67). $\dfrac{44}{45} - \dfrac{7}{9} - \dfrac{1}{15} =$

68). $\dfrac{52}{54} - \dfrac{4}{6} - \dfrac{2}{27} =$

69). $\dfrac{59}{60} - \dfrac{1}{4} - \dfrac{2}{30} =$

70). $\dfrac{39}{40} - \dfrac{6}{10} - \dfrac{1}{20} =$

71). $\dfrac{54}{56} - \dfrac{5}{8} - \dfrac{2}{28} =$

72). $\dfrac{22}{24} - \dfrac{2}{8} - \dfrac{1}{12} =$

73). $\dfrac{48}{50} - \dfrac{2}{10} - \dfrac{2}{25} =$

74). $\dfrac{43}{45} - \dfrac{2}{5} - \dfrac{1}{15} =$

75). $\dfrac{46}{48} - \dfrac{8}{12} - \dfrac{2}{24} =$

76). $\dfrac{29}{30} - \dfrac{8}{10} - \dfrac{1}{15} =$

77). $\dfrac{53}{54} - \dfrac{11}{18} - \dfrac{1}{27} =$

78). $\dfrac{44}{45} - \dfrac{6}{9} - \dfrac{1}{15} =$

79). $\dfrac{18}{20} - \dfrac{3}{5} - \dfrac{1}{10} =$

80). $\dfrac{30}{32} - \dfrac{2}{4} - \dfrac{2}{16} =$

81). $\dfrac{28}{30} - \dfrac{1}{3} - \dfrac{1}{15} =$

82). $\dfrac{44}{45} - \dfrac{1}{5} - \dfrac{1}{15} =$

83). $\dfrac{54}{56} - \dfrac{5}{7} - \dfrac{2}{28} =$

84). $\dfrac{58}{60} - \dfrac{7}{12} - \dfrac{1}{30} =$

85). $\dfrac{28}{30} - \dfrac{3}{5} - \dfrac{2}{15} =$

86). $\dfrac{43}{45} - \dfrac{1}{5} - \dfrac{2}{15} =$

87). $\dfrac{49}{50} - \dfrac{7}{10} - \dfrac{2}{25} =$

88). $\dfrac{39}{40} - \dfrac{2}{10} - \dfrac{1}{20} =$

89). $\dfrac{52}{54} - \dfrac{1}{3} - \dfrac{2}{27} =$

90). $\dfrac{27}{28} - \dfrac{1}{4} - \dfrac{2}{14} =$

91). $\dfrac{44}{45} - \dfrac{1}{9} - \dfrac{2}{15} =$

92). $\dfrac{53}{54} - \dfrac{1}{9} - \dfrac{1}{27} =$

93). $\dfrac{54}{56} - \dfrac{2}{7} - \dfrac{1}{28} =$

94). $\dfrac{43}{45} - \dfrac{5}{9} - \dfrac{2}{15} =$

95). $\dfrac{27}{28} - \dfrac{2}{4} - \dfrac{1}{14} =$

96). $\dfrac{40}{42} - \dfrac{4}{14} - \dfrac{2}{21} =$

97). $\dfrac{19}{20} - \dfrac{2}{5} - \dfrac{2}{10} =$

98). $\dfrac{29}{30} - \dfrac{1}{3} - \dfrac{2}{15} =$

99). $\dfrac{47}{48} - \dfrac{3}{6} - \dfrac{1}{24} =$

100). $\dfrac{26}{28} - \dfrac{4}{7} - \dfrac{1}{14} =$

ANSWER

1). $\dfrac{7}{8} - \dfrac{5}{8} - \dfrac{1}{8} = \dfrac{1}{8}$

2). $\dfrac{7}{8} - \dfrac{2}{8} - \dfrac{1}{8} = \dfrac{4}{8}$

3). $\dfrac{10}{11} - \dfrac{4}{11} - \dfrac{5}{11} = \dfrac{1}{11}$

4). $\dfrac{10}{11} - \dfrac{4}{11} - \dfrac{1}{11} = \dfrac{5}{11}$

5). $\dfrac{5}{6} - \dfrac{3}{6} - \dfrac{1}{6} = \dfrac{1}{6}$

6). $\dfrac{10}{11} - \dfrac{7}{11} - \dfrac{2}{11} = \dfrac{1}{11}$

7). $\dfrac{6}{7} - \dfrac{1}{7} - \dfrac{3}{7} = \dfrac{2}{7}$

8). $\dfrac{5}{6} - \dfrac{2}{6} - \dfrac{2}{6} = \dfrac{1}{6}$

9). $\dfrac{7}{8} - \dfrac{3}{8} - \dfrac{1}{8} = \dfrac{3}{8}$

10). $\dfrac{5}{6} - \dfrac{3}{6} - \dfrac{1}{6} = \dfrac{1}{6}$

11). $\dfrac{11}{12} - \dfrac{6}{12} - \dfrac{3}{12} = \dfrac{2}{12}$

12). $\dfrac{9}{10} - \dfrac{1}{10} - \dfrac{6}{10} = \dfrac{2}{10}$

13). $\dfrac{6}{7} - \dfrac{4}{7} - \dfrac{1}{7} = \dfrac{1}{7}$

14). $\dfrac{8}{9} - \dfrac{5}{9} - \dfrac{1}{9} = \dfrac{2}{9}$

15). $\dfrac{9}{10} - \dfrac{5}{10} - \dfrac{1}{10} = \dfrac{3}{10}$

16). $\dfrac{8}{9} - \dfrac{1}{9} - \dfrac{4}{9} = \dfrac{3}{9}$

17). $\dfrac{5}{6} - \dfrac{3}{6} - \dfrac{1}{6} = \dfrac{1}{6}$

18). $\dfrac{9}{10} - \dfrac{5}{10} - \dfrac{3}{10} = \dfrac{1}{10}$

19). $\dfrac{9}{10} - \dfrac{5}{10} - \dfrac{1}{10} = \dfrac{3}{10}$

20). $\dfrac{7}{8} - \dfrac{5}{8} - \dfrac{1}{8} = \dfrac{1}{8}$

21). $\frac{8}{10} - \frac{1}{2} - \frac{3}{10} = 0$

22). $\frac{4}{5} - \frac{1}{5} - \frac{1}{5} = \frac{2}{5}$

23). $\frac{8}{10} - \frac{1}{3} - \frac{3}{10} = \frac{5}{30}$

24). $\frac{4}{5} - \frac{2}{5} - \frac{3}{10} = \frac{1}{10}$

25). $\frac{9}{10} - \frac{1}{4} - \frac{1}{5} = \frac{9}{20}$

26). $\frac{4}{5} - \frac{1}{4} - \frac{1}{10} = \frac{9}{20}$

27). $\frac{4}{5} - \frac{1}{2} - \frac{1}{5} = \frac{1}{10}$

28). $\frac{8}{10} - \frac{1}{4} - \frac{3}{10} = \frac{5}{20}$

29). $\frac{9}{10} - \frac{1}{5} - \frac{2}{10} = \frac{5}{10}$

30). $\frac{9}{10} - \frac{1}{5} - \frac{2}{10} = \frac{5}{10}$

31). $\frac{8}{10} - \frac{1}{5} - \frac{1}{10} = \frac{5}{10}$

32). $\frac{9}{10} - \frac{1}{2} - \frac{1}{5} = \frac{2}{10}$

33). $\frac{9}{10} - \frac{1}{5} - \frac{1}{5} = \frac{5}{10}$

34). $\frac{4}{5} - \frac{1}{4} - \frac{1}{10} = \frac{9}{20}$

35). $\frac{4}{5} - \frac{1}{3} - \frac{1}{10} = \frac{11}{30}$

36). $\frac{4}{5} - \frac{1}{3} - \frac{1}{10} = \frac{1}{10}$

37). $\frac{4}{5} - \frac{1}{2} - \frac{1}{5} = \frac{1}{10}$

38). $\frac{4}{5} - \frac{1}{2} - \frac{1}{5} = \frac{5}{30}$

39). $\frac{4}{5} - \frac{1}{3} - \frac{3}{10} = \frac{1}{10}$

40). $\frac{9}{10} - \frac{1}{2} - \frac{3}{10} = \frac{1}{10}$

41). $\frac{11}{12} - \frac{1}{3} - \frac{1}{6} = \frac{5}{12}$

42). $\frac{11}{12} - \frac{1}{3} - \frac{1}{6} = \frac{5}{12}$

43). $\frac{11}{12} - \frac{2}{4} - \frac{1}{6} = \frac{3}{12}$

44). $\frac{23}{24} - \frac{2}{4} - \frac{1}{12} = \frac{9}{24}$

45). $\frac{26}{28} - \frac{4}{7} - \frac{1}{14} = \frac{8}{28}$

46). $\frac{18}{20} - \frac{3}{5} - \frac{1}{10} = \frac{4}{20}$

47). $\frac{15}{16} - \frac{1}{4} - \frac{2}{8} = \frac{7}{16}$

48). $\frac{22}{24} - \frac{2}{8} - \frac{2}{12} = \frac{12}{24}$

49). $\frac{26}{28} - \frac{1}{4} - \frac{1}{14} = \frac{17}{28}$

50). $\frac{11}{12} - \frac{1}{3} - \frac{1}{6} = \frac{5}{12}$

51). $\frac{26}{28} - \frac{2}{7} - \frac{1}{14} = \frac{16}{28}$

52). $\frac{11}{12} - \frac{1}{3} - \frac{1}{6} = \frac{5}{12}$

53). $\frac{16}{18} - \frac{1}{3} - \frac{1}{9} = \frac{8}{18}$

54). $\frac{22}{24} - \frac{4}{8} - \frac{2}{12} = \frac{6}{24}$

55). $\frac{26}{28} - \frac{1}{4} - \frac{2}{14} = \frac{15}{28}$

56). $\frac{28}{30} - \frac{4}{6} - \frac{2}{15} = \frac{4}{30}$

57). $\frac{18}{20} - \frac{1}{4} - \frac{2}{10} = \frac{9}{20}$

58). $\frac{29}{30} - \frac{1}{6} - \frac{1}{15} = \frac{22}{30}$

59). $\frac{26}{28} - \frac{2}{7} - \frac{2}{14} = \frac{14}{28}$

60). $\frac{29}{30} - \frac{4}{6} - \frac{1}{15} = \frac{7}{30}$

61). $\dfrac{40}{42} - \dfrac{3}{6} - \dfrac{2}{21} = \dfrac{15}{42}$ 71). $\dfrac{54}{56} - \dfrac{5}{8} - \dfrac{2}{28} = \dfrac{15}{56}$

62). $\dfrac{49}{50} - \dfrac{5}{10} - \dfrac{1}{25} = \dfrac{22}{50}$ 72). $\dfrac{22}{24} - \dfrac{2}{8} - \dfrac{1}{12} = \dfrac{14}{24}$

63). $\dfrac{44}{45} - \dfrac{3}{5} - \dfrac{2}{15} = \dfrac{11}{45}$ 73). $\dfrac{48}{50} - \dfrac{2}{10} - \dfrac{2}{25} = \dfrac{34}{50}$

64). $\dfrac{59}{60} - \dfrac{1}{6} - \dfrac{1}{30} = \dfrac{47}{60}$ 74). $\dfrac{43}{45} - \dfrac{2}{5} - \dfrac{1}{15} = \dfrac{22}{45}$

65). $\dfrac{49}{50} - \dfrac{1}{5} - \dfrac{1}{25} = \dfrac{37}{50}$ 75). $\dfrac{46}{48} - \dfrac{8}{12} - \dfrac{2}{24} = \dfrac{10}{48}$

66). $\dfrac{39}{40} - \dfrac{2}{4} - \dfrac{1}{20} = \dfrac{17}{40}$ 76). $\dfrac{29}{30} - \dfrac{8}{10} - \dfrac{1}{15} = \dfrac{3}{30}$

67). $\dfrac{44}{45} - \dfrac{7}{9} - \dfrac{1}{15} = \dfrac{6}{45}$ 77). $\dfrac{53}{54} - \dfrac{11}{18} - \dfrac{1}{27} = \dfrac{18}{54}$

68). $\dfrac{52}{54} - \dfrac{4}{6} - \dfrac{2}{27} = \dfrac{12}{54}$ 78). $\dfrac{44}{45} - \dfrac{6}{9} - \dfrac{1}{15} = \dfrac{11}{45}$

69). $\dfrac{59}{60} - \dfrac{1}{4} - \dfrac{2}{30} = \dfrac{40}{60}$ 79). $\dfrac{18}{20} - \dfrac{3}{5} - \dfrac{1}{10} = \dfrac{4}{20}$

70). $\dfrac{39}{40} - \dfrac{6}{10} - \dfrac{1}{20} = \dfrac{13}{40}$ 80). $\dfrac{30}{32} - \dfrac{2}{4} - \dfrac{2}{16} = \dfrac{10}{32}$

81). $\frac{28}{30} - \frac{1}{3} - \frac{1}{15} = \frac{16}{30}$

82). $\frac{44}{45} - \frac{1}{5} - \frac{1}{15} = \frac{32}{45}$

83). $\frac{54}{56} - \frac{5}{7} - \frac{2}{28} = \frac{10}{56}$

84). $\frac{58}{60} - \frac{7}{12} - \frac{1}{30} = \frac{21}{60}$

85). $\frac{28}{30} - \frac{3}{5} - \frac{2}{15} = \frac{6}{30}$

86). $\frac{43}{45} - \frac{1}{5} - \frac{2}{15} = \frac{28}{45}$

87). $\frac{49}{50} - \frac{7}{10} - \frac{2}{25} = \frac{10}{50}$

88). $\frac{39}{40} - \frac{2}{10} - \frac{1}{20} = \frac{29}{40}$

89). $\frac{52}{54} - \frac{1}{3} - \frac{2}{27} = \frac{30}{54}$

90). $\frac{27}{28} - \frac{1}{4} - \frac{2}{14} = \frac{16}{28}$

91). $\frac{44}{45} - \frac{1}{9} - \frac{2}{15} = \frac{33}{45}$

92). $\frac{53}{54} - \frac{1}{9} - \frac{1}{27} = \frac{45}{54}$

93). $\frac{54}{56} - \frac{2}{7} - \frac{1}{28} = \frac{36}{56}$

94). $\frac{43}{45} - \frac{5}{9} - \frac{2}{15} = \frac{12}{45}$

95). $\frac{27}{28} - \frac{2}{4} - \frac{1}{14} = \frac{11}{28}$

96). $\frac{40}{42} - \frac{4}{14} - \frac{2}{21} = \frac{24}{42}$

97). $\frac{19}{20} - \frac{2}{5} - \frac{2}{10} = \frac{7}{20}$

98). $\frac{29}{30} - \frac{1}{3} - \frac{2}{15} = \frac{15}{30}$

99). $\frac{47}{48} - \frac{3}{6} - \frac{1}{24} = \frac{21}{48}$

100). $\frac{26}{28} - \frac{4}{7} - \frac{1}{14} = \frac{8}{28}$

SUBTRACTION MIXED FRACTION

1). $6\frac{8}{10} - 2\frac{1}{2} =$

2). $9\frac{4}{5} - 3\frac{1}{2} =$

3). $7\frac{1}{3} - 2\frac{1}{4} =$

4). $6\frac{9}{10} - 4\frac{1}{4} =$

5). $5\frac{1}{3} - 1\frac{2}{10} =$

6). $5\frac{1}{2} - 4\frac{3}{10} =$

7). $7\frac{2}{3} - 3\frac{1}{5} =$

8). $5\frac{1}{2} - 4\frac{1}{4} =$

9). $6\frac{2}{3} - 1\frac{1}{2} =$

10). $5\frac{4}{5} - 2\frac{3}{4} =$

11). $7\frac{2}{3} - 4\frac{1}{2} =$

12). $7\frac{3}{5} - 2\frac{1}{3} =$

13). $9\frac{1}{4} - 2\frac{1}{5} =$

14). $6\frac{7}{10} - 1\frac{1}{2} =$

15). $8\frac{1}{2} - 1\frac{1}{3} =$

16). $8\frac{3}{5} - 2\frac{1}{3} =$

17). $5\frac{2}{3} - 3\frac{4}{10} =$

18). $8\frac{3}{4} - 1\frac{1}{2} =$

19). $8\frac{7}{10} - 2\frac{2}{3} =$

20). $7\frac{3}{4} - 2\frac{6}{10} =$

21). $7\frac{9}{26} - 2\frac{2}{13} =$

22). $9\frac{4}{9} - 4\frac{1}{3} =$

23). $8\frac{3}{7} - 1\frac{4}{21} =$

24). $8\frac{1}{3} - 2\frac{2}{30} =$

25). $6\frac{3}{4} - 4\frac{2}{10} =$

26). $8\frac{12}{14} - 1\frac{3}{7} =$

27). $6\frac{7}{11} - 1\frac{6}{22} =$

28). $7\frac{7}{11} - 2\frac{11}{22} =$

29). $9\frac{5}{9} - 4\frac{5}{27} =$

30). $9\frac{11}{14} - 3\frac{5}{7} =$

31). $5\frac{2}{3} - 1\frac{2}{5} =$

32). $5\frac{2}{7} - 3\frac{2}{28} =$

33). $5\frac{7}{8} - 3\frac{3}{16} =$

34). $7\frac{1}{3} - 2\frac{2}{12} =$

35). $9\frac{4}{5} - 3\frac{7}{10} =$

36). $5\frac{6}{10} - 1\frac{2}{20} =$

37). $5\frac{5}{9} - 4\frac{11}{27} =$

38). $7\frac{6}{7} - 3\frac{11}{14} =$

39). $7\frac{2}{3} - 3\frac{11}{24} =$

40). $5\frac{3}{6} - 2\frac{2}{12} =$

ANSWER

1). $6\frac{8}{10} - 2\frac{1}{2} = 4\frac{3}{10}$

2). $9\frac{4}{5} - 3\frac{1}{2} = 6\frac{3}{10}$

3). $7\frac{1}{3} - 2\frac{1}{4} = 5\frac{1}{12}$

4). $6\frac{9}{10} - 4\frac{1}{4} = 2\frac{13}{20}$

5). $5\frac{1}{3} - 1\frac{2}{10} = 4\frac{4}{30}$

6). $5\frac{1}{2} - 4\frac{3}{10} = 1\frac{2}{10}$

7). $7\frac{2}{3} - 3\frac{1}{5} = 4\frac{7}{15}$

8). $5\frac{1}{2} - 4\frac{1}{4} = 1\frac{1}{4}$

9). $6\frac{2}{3} - 1\frac{1}{2} = 5\frac{1}{6}$

10). $5\frac{4}{5} - 2\frac{3}{4} = 3\frac{1}{20}$

11). $7\frac{2}{3} - 4\frac{1}{2} = 3\frac{1}{6}$

12). $7\frac{3}{5} - 2\frac{1}{3} = 5\frac{4}{15}$

13). $9\frac{1}{4} - 2\frac{1}{5} = 7\frac{1}{20}$

14). $6\frac{7}{10} - 1\frac{1}{2} = 5\frac{2}{10}$

15). $8\frac{1}{2} - 1\frac{1}{3} = 7\frac{1}{6}$

16). $8\frac{3}{5} - 2\frac{1}{3} = 6\frac{4}{15}$

17). $5\frac{2}{3} - 3\frac{4}{10} = 2\frac{8}{30}$

18). $8\frac{3}{4} - 1\frac{1}{2} = 7\frac{1}{4}$

19). $8\frac{7}{10} - 2\frac{2}{3} = 6\frac{1}{30}$

20). $7\frac{3}{4} - 2\frac{6}{10} = 5\frac{3}{20}$

21). $7\frac{9}{26} - 2\frac{2}{13} = 5\frac{5}{26}$ 31). $5\frac{2}{3} - 1\frac{2}{5} = 4\frac{4}{15}$

22). $9\frac{4}{9} - 4\frac{1}{3} = 5\frac{1}{9}$ 32). $5\frac{2}{7} - 3\frac{2}{28} = 2\frac{6}{28}$

23). $8\frac{3}{7} - 1\frac{4}{21} = 7\frac{5}{21}$ 33). $5\frac{7}{8} - 3\frac{3}{16} = 2\frac{11}{16}$

24). $8\frac{1}{3} - 2\frac{2}{30} = 6\frac{8}{30}$ 34). $7\frac{1}{3} - 2\frac{2}{12} = 5\frac{2}{12}$

25). $6\frac{3}{4} - 4\frac{2}{10} = 2\frac{11}{20}$ 35). $9\frac{4}{5} - 3\frac{7}{10} = 6\frac{1}{10}$

26). $8\frac{12}{14} - 1\frac{3}{7} = 7\frac{6}{14}$ 36). $5\frac{6}{10} - 1\frac{2}{20} = 4\frac{10}{20}$

27). $6\frac{7}{11} - 1\frac{6}{22} = 5\frac{8}{22}$ 37). $5\frac{5}{9} - 4\frac{11}{27} = 1\frac{4}{27}$

28). $7\frac{7}{11} - 2\frac{11}{22} = 5\frac{3}{22}$ 38). $7\frac{6}{7} - 3\frac{11}{14} = 4\frac{1}{14}$

29). $9\frac{5}{9} - 4\frac{5}{27} = 5\frac{10}{27}$ 39). $7\frac{2}{3} - 3\frac{11}{24} = 4\frac{5}{24}$

30). $9\frac{11}{14} - 3\frac{5}{7} = 6\frac{1}{14}$ 40). $5\frac{3}{6} - 2\frac{2}{12} = 3\frac{4}{12}$

MULTIPLICATION OF FRACTIONS

1). $\frac{2}{4} \times \frac{2}{3} =$

2). $\frac{3}{4} \times \frac{1}{2} =$

3). $\frac{2}{4} \times \frac{1}{2} =$

4). $\frac{4}{5} \times \frac{1}{4} =$

5). $\frac{4}{10} \times \frac{1}{2} =$

6). $\frac{4}{10} \times \frac{1}{4} =$

7). $\frac{1}{4} \times \frac{1}{3} =$

8). $\frac{3}{4} \times \frac{3}{5} =$

9). $\frac{1}{2} \times \frac{7}{10} =$

10). $\frac{1}{5} \times \frac{3}{4} =$

11). $\frac{2}{3} \times \frac{1}{4} =$

12). $\frac{1}{3} \times \frac{3}{4} =$

13). $\frac{6}{10} \times \frac{1}{2} =$

14). $\frac{1}{2} \times \frac{3}{10} =$

15). $\frac{1}{2} \times \frac{2}{3} =$

16). $\frac{1}{2} \times \frac{2}{5} =$

17). $\frac{3}{10} \times \frac{3}{5} =$

18). $\frac{1}{10} \times \frac{1}{2} =$

19). $\frac{1}{5} \times \frac{2}{4} =$

20). $\frac{7}{10} \times \frac{4}{5} =$

21). $\frac{1}{3} \times \frac{1}{6} =$

22). $\frac{3}{7} \times \frac{2}{3} =$

23). $\frac{1}{4} \times \frac{3}{6} =$

24). $\frac{2}{5} \times \frac{1}{10} =$

25). $\frac{7}{9} \times \frac{4}{7} =$

26). $\frac{1}{5} \times \frac{8}{9} =$

27). $\frac{9}{10} \times \frac{1}{9} =$

28). $\frac{3}{10} \times \frac{5}{6} =$

29). $\frac{2}{6} \times \frac{1}{5} =$

30). $\frac{2}{10} \times \frac{2}{6} =$

31). $\frac{7}{8} \times \frac{4}{6} =$

32). $\frac{2}{5} \times \frac{1}{2} =$

33). $\frac{3}{4} \times \frac{1}{2} =$

34). $\frac{1}{2} \times \frac{1}{3} =$

35). $\frac{7}{9} \times \frac{5}{7} =$

36). $\frac{1}{6} \times \frac{2}{7} =$

37). $\frac{1}{2} \times \frac{6}{8} =$

38). $\frac{2}{3} \times \frac{3}{4} =$

39). $\frac{5}{9} \times \frac{5}{6} =$

40). $\frac{5}{7} \times \frac{1}{2} =$

41). $\frac{1}{3} \times \frac{12}{14} =$

42). $\frac{2}{3} \times \frac{5}{10} =$

43). $\frac{4}{5} \times \frac{1}{3} =$

44). $\frac{2}{15} \times \frac{8}{18} =$

45). $\frac{2}{6} \times \frac{1}{4} =$

46). $\frac{2}{8} \times \frac{1}{2} =$

47). $\frac{2}{6} \times \frac{11}{16} =$

48). $\frac{2}{3} \times \frac{2}{12} =$

49). $\frac{8}{14} \times \frac{4}{5} =$

50). $\frac{11}{12} \times \frac{3}{4} =$

51). $\frac{17}{18} \times \frac{8}{15} =$

52). $\frac{1}{4} \times \frac{17}{20} =$

53). $\frac{4}{5} \times \frac{6}{15} =$

54). $\frac{13}{16} \times \frac{3}{14} =$

55). $\frac{7}{15} \times \frac{9}{16} =$

56). $\frac{7}{14} \times \frac{5}{10} =$

57). $\frac{2}{14} \times \frac{5}{6} =$

58). $\frac{3}{9} \times \frac{10}{18} =$

59). $\frac{9}{15} \times \frac{6}{7} =$

60). $\frac{3}{4} \times \frac{4}{6} =$

61). $\dfrac{2}{4} \times \dfrac{1}{2} =$

62). $\dfrac{1}{4} \times \dfrac{2}{3} =$

63). $\dfrac{5}{10} \times \dfrac{1}{3} =$

64). $\dfrac{2}{10} \times \dfrac{1}{2} =$

65). $\dfrac{1}{3} \times \dfrac{4}{5} =$

66). $\dfrac{1}{10} \times \dfrac{3}{4} =$

67). $\dfrac{1}{2} \times \dfrac{3}{4} =$

68). $\dfrac{8}{10} \times \dfrac{3}{5} =$

69). $\dfrac{2}{4} \times \dfrac{2}{3} =$

70). $\dfrac{3}{5} \times \dfrac{2}{4} =$

71). $\dfrac{3}{9} \times \dfrac{2}{5} =$

72). $\dfrac{5}{8} \times \dfrac{3}{4} =$

73). $\dfrac{2}{6} \times \dfrac{1}{2} =$

74). $\dfrac{5}{6} \times \dfrac{1}{8} =$

75). $\dfrac{6}{10} \times \dfrac{4}{7} =$

76). $\dfrac{4}{6} \times \dfrac{3}{4} =$

77). $\dfrac{1}{2} \times \dfrac{7}{10} =$

78). $\dfrac{5}{9} \times \dfrac{2}{3} =$

79). $\dfrac{1}{6} \times \dfrac{3}{5} =$

80). $\dfrac{4}{6} \times \dfrac{8}{9} =$

81). $\dfrac{5}{12} \times \dfrac{2}{3} =$

82). $\dfrac{11}{12} \times \dfrac{15}{16} =$

83). $\dfrac{4}{5} \times \dfrac{1}{2} =$

84). $\dfrac{6}{16} \times \dfrac{5}{8} =$

85). $\dfrac{4}{14} \times \dfrac{5}{8} =$

86). $\dfrac{1}{2} \times \dfrac{7}{9} =$

87). $\dfrac{3}{15} \times \dfrac{1}{6} =$

88). $\dfrac{7}{14} \times \dfrac{4}{6} =$

89). $\dfrac{1}{12} \times \dfrac{4}{5} =$

90). $\dfrac{2}{3} \times \dfrac{6}{18} =$

91). $\dfrac{6}{10} \times \dfrac{16}{25} =$

92). $\dfrac{8}{14} \times \dfrac{2}{4} =$

93). $\dfrac{3}{12} \times \dfrac{5}{14} =$

94). $\dfrac{4}{15} \times \dfrac{1}{2} =$

95). $\dfrac{9}{22} \times \dfrac{13}{14} =$

96). $\dfrac{7}{14} \times \dfrac{5}{20} =$

97). $\dfrac{5}{14} \times \dfrac{5}{10} =$

98). $\dfrac{4}{5} \times \dfrac{3}{12} =$

99). $\dfrac{2}{14} \times \dfrac{2}{3} =$

100). $\dfrac{9}{15} \times \dfrac{2}{4} =$

ANSWER

1). $\dfrac{2}{4} \times \dfrac{2}{3} = \dfrac{4}{12}$

2). $\dfrac{3}{4} \times \dfrac{1}{2} = \dfrac{3}{8}$

3). $\dfrac{2}{4} \times \dfrac{1}{2} = \dfrac{2}{8}$

4). $\dfrac{4}{5} \times \dfrac{1}{4} = \dfrac{4}{20}$

5). $\dfrac{4}{10} \times \dfrac{1}{2} = \dfrac{4}{20}$

6). $\dfrac{4}{10} \times \dfrac{1}{4} = \dfrac{4}{40}$

7). $\dfrac{1}{4} \times \dfrac{1}{3} = \dfrac{1}{12}$

8). $\dfrac{3}{4} \times \dfrac{3}{5} = \dfrac{9}{20}$

9). $\dfrac{1}{2} \times \dfrac{7}{10} = \dfrac{7}{20}$

10). $\dfrac{1}{5} \times \dfrac{3}{4} = \dfrac{3}{20}$

11). $\dfrac{2}{3} \times \dfrac{1}{4} = \dfrac{2}{12}$

12). $\dfrac{1}{3} \times \dfrac{3}{4} = \dfrac{3}{12}$

13). $\dfrac{6}{10} \times \dfrac{1}{2} = \dfrac{6}{20}$

14). $\dfrac{1}{2} \times \dfrac{3}{10} = \dfrac{3}{20}$

15). $\dfrac{1}{2} \times \dfrac{2}{3} = \dfrac{2}{6}$

16). $\dfrac{1}{2} \times \dfrac{2}{5} = \dfrac{2}{10}$

17). $\dfrac{3}{10} \times \dfrac{3}{5} = \dfrac{9}{50}$

18). $\dfrac{1}{10} \times \dfrac{1}{2} = \dfrac{1}{20}$

19). $\dfrac{1}{5} \times \dfrac{2}{4} = \dfrac{2}{20}$

20). $\dfrac{7}{10} \times \dfrac{4}{5} = \dfrac{28}{50}$

21). $\dfrac{1}{3} \times \dfrac{1}{6} = \dfrac{1}{18}$

22). $\dfrac{3}{7} \times \dfrac{2}{3} = \dfrac{6}{21}$

23). $\dfrac{1}{4} \times \dfrac{3}{6} = \dfrac{3}{24}$

24). $\dfrac{2}{5} \times \dfrac{1}{10} = \dfrac{2}{50}$

25). $\dfrac{7}{9} \times \dfrac{4}{7} = \dfrac{28}{63}$

26). $\dfrac{1}{5} \times \dfrac{8}{9} = \dfrac{8}{45}$

27). $\dfrac{9}{10} \times \dfrac{1}{9} = \dfrac{9}{90}$

28). $\dfrac{3}{10} \times \dfrac{5}{6} = \dfrac{15}{60}$

29). $\dfrac{2}{6} \times \dfrac{1}{5} = \dfrac{2}{30}$

30). $\dfrac{2}{10} \times \dfrac{2}{6} = \dfrac{4}{60}$

31). $\dfrac{7}{8} \times \dfrac{4}{6} = \dfrac{28}{48}$

32). $\dfrac{2}{5} \times \dfrac{1}{2} = \dfrac{2}{10}$

33). $\dfrac{3}{4} \times \dfrac{1}{2} = \dfrac{3}{8}$

34). $\dfrac{1}{2} \times \dfrac{1}{3} = \dfrac{1}{6}$

35). $\dfrac{7}{9} \times \dfrac{5}{7} = \dfrac{35}{63}$

36). $\dfrac{1}{6} \times \dfrac{2}{7} = \dfrac{2}{42}$

37). $\dfrac{1}{2} \times \dfrac{6}{8} = \dfrac{6}{16}$

38). $\dfrac{2}{3} \times \dfrac{3}{4} = \dfrac{6}{12}$

39). $\dfrac{5}{9} \times \dfrac{5}{6} = \dfrac{25}{54}$

40). $\dfrac{5}{7} \times \dfrac{1}{2} = \dfrac{5}{14}$

41). $\dfrac{1}{3} \times \dfrac{12}{14} = \dfrac{12}{42}$

42). $\dfrac{2}{3} \times \dfrac{5}{10} = \dfrac{10}{30}$

43). $\dfrac{4}{5} \times \dfrac{1}{3} = \dfrac{4}{15}$

44). $\dfrac{2}{15} \times \dfrac{8}{18} = \dfrac{16}{270}$

45). $\dfrac{2}{6} \times \dfrac{1}{4} = \dfrac{2}{24}$

46). $\dfrac{2}{8} \times \dfrac{1}{2} = \dfrac{2}{16}$

47). $\dfrac{2}{6} \times \dfrac{11}{16} = \dfrac{22}{96}$

48). $\dfrac{2}{3} \times \dfrac{2}{12} = \dfrac{4}{36}$

49). $\dfrac{8}{14} \times \dfrac{4}{5} = \dfrac{32}{70}$

50). $\dfrac{11}{12} \times \dfrac{3}{4} = \dfrac{33}{48}$

51). $\dfrac{17}{18} \times \dfrac{8}{15} = \dfrac{136}{270}$

52). $\dfrac{1}{4} \times \dfrac{17}{20} = \dfrac{17}{80}$

53). $\dfrac{4}{5} \times \dfrac{6}{15} = \dfrac{24}{75}$

54). $\dfrac{13}{16} \times \dfrac{3}{14} = \dfrac{39}{224}$

55). $\dfrac{7}{15} \times \dfrac{9}{16} = \dfrac{63}{240}$

56). $\dfrac{7}{14} \times \dfrac{5}{10} = \dfrac{35}{140}$

57). $\dfrac{2}{14} \times \dfrac{5}{6} = \dfrac{10}{84}$

58). $\dfrac{3}{9} \times \dfrac{10}{18} = \dfrac{30}{162}$

59). $\dfrac{9}{15} \times \dfrac{6}{7} = \dfrac{54}{105}$

60). $\dfrac{3}{4} \times \dfrac{4}{6} = \dfrac{12}{24}$

61). $\frac{2}{4} \times \frac{1}{2} = \frac{1}{4}$

62). $\frac{1}{4} \times \frac{2}{3} = \frac{1}{6}$

63). $\frac{5}{10} \times \frac{1}{3} = \frac{1}{6}$

64). $\frac{2}{10} \times \frac{1}{2} = \frac{1}{10}$

65). $\frac{1}{3} \times \frac{4}{5} = \frac{4}{15}$

66). $\frac{1}{10} \times \frac{3}{4} = \frac{3}{40}$

67). $\frac{1}{2} \times \frac{3}{4} = \frac{3}{8}$

68). $\frac{8}{10} \times \frac{3}{5} = \frac{12}{25}$

69). $\frac{2}{4} \times \frac{2}{3} = \frac{1}{3}$

70). $\frac{3}{5} \times \frac{2}{4} = \frac{3}{10}$

71). $\frac{3}{9} \times \frac{2}{5} = \frac{2}{15}$

72). $\frac{5}{8} \times \frac{3}{4} = \frac{15}{32}$

73). $\frac{2}{6} \times \frac{1}{2} = \frac{1}{6}$

74). $\frac{5}{6} \times \frac{1}{8} = \frac{5}{48}$

75). $\frac{6}{10} \times \frac{4}{7} = \frac{12}{35}$

76). $\frac{4}{6} \times \frac{3}{4} = \frac{1}{2}$

77). $\frac{1}{2} \times \frac{7}{10} = \frac{7}{20}$

78). $\frac{5}{9} \times \frac{2}{3} = \frac{10}{27}$

79). $\frac{1}{6} \times \frac{3}{5} = \frac{1}{10}$

80). $\frac{4}{6} \times \frac{8}{9} = \frac{16}{27}$

81). $\dfrac{5}{12} \times \dfrac{2}{3} = \dfrac{5}{18}$

82). $\dfrac{11}{12} \times \dfrac{15}{16} = \dfrac{55}{64}$

83). $\dfrac{4}{5} \times \dfrac{1}{2} = \dfrac{2}{5}$

84). $\dfrac{6}{16} \times \dfrac{5}{8} = \dfrac{15}{64}$

85). $\dfrac{4}{14} \times \dfrac{5}{8} = \dfrac{5}{28}$

86). $\dfrac{1}{2} \times \dfrac{7}{9} = \dfrac{7}{18}$

87). $\dfrac{3}{15} \times \dfrac{1}{6} = \dfrac{1}{30}$

88). $\dfrac{7}{14} \times \dfrac{4}{6} = \dfrac{1}{3}$

89). $\dfrac{1}{12} \times \dfrac{4}{5} = \dfrac{1}{15}$

90). $\dfrac{2}{3} \times \dfrac{6}{18} = \dfrac{2}{9}$

91). $\dfrac{6}{10} \times \dfrac{16}{25} = \dfrac{48}{125}$

92). $\dfrac{8}{14} \times \dfrac{2}{4} = \dfrac{2}{7}$

93). $\dfrac{3}{12} \times \dfrac{5}{14} = \dfrac{5}{56}$

94). $\dfrac{4}{15} \times \dfrac{1}{2} = \dfrac{2}{15}$

95). $\dfrac{9}{22} \times \dfrac{13}{14} = \dfrac{117}{308}$

96). $\dfrac{7}{14} \times \dfrac{5}{20} = \dfrac{1}{8}$

97). $\dfrac{5}{14} \times \dfrac{5}{10} = \dfrac{5}{28}$

98). $\dfrac{4}{5} \times \dfrac{3}{12} = \dfrac{1}{5}$

99). $\dfrac{2}{14} \times \dfrac{2}{3} = \dfrac{2}{21}$

100). $\dfrac{9}{15} \times \dfrac{2}{4} = \dfrac{3}{10}$

MULTIPLICATION MIXED FRACTION

1). $2\frac{2}{3} \times 4\frac{2}{5} =$

2). $4\frac{3}{4} \times 3\frac{1}{2} =$

3). $2\frac{4}{5} \times 3\frac{2}{3} =$

4). $3\frac{3}{5} \times 3\frac{1}{2} =$

5). $3\frac{1}{3} \times 3\frac{1}{2} =$

6). $4\frac{1}{3} \times 2\frac{1}{2} =$

7). $2\frac{2}{5} \times 2\frac{1}{2} =$

8). $3\frac{1}{2} \times 3\frac{1}{4} =$

9). $4\frac{4}{5} \times 4\frac{1}{3} =$

10). $4\frac{1}{2} \times 4\frac{4}{5} =$

11). $2\frac{1}{2} \times 4\frac{1}{10} =$

12). $2\frac{2}{3} \times 2\frac{1}{5} =$

13). $4\frac{1}{5} \times 3\frac{1}{2} =$

14). $3\frac{1}{2} \times 4\frac{2}{5} =$

15). $3\frac{1}{4} \times 3\frac{1}{2} =$

16). $4\frac{1}{4} \times 2\frac{2}{3} =$

17). $2\frac{1}{4} \times 3\frac{3}{5} =$

18). $2\frac{1}{3} \times 4\frac{1}{2} =$

19). $2\frac{9}{10} \times 4\frac{2}{5} =$

20). $2\frac{4}{5} \times 3\frac{3}{5} =$

21). $2\frac{1}{2} \times 4\frac{1}{10} =$

22). $2\frac{2}{3} \times 2\frac{1}{5} =$

23). $4\frac{1}{5} \times 3\frac{1}{2} =$

24). $3\frac{1}{2} \times 4\frac{2}{5} =$

25). $3\frac{1}{4} \times 3\frac{1}{2} =$

26). $4\frac{1}{4} \times 2\frac{2}{3} =$

27). $2\frac{1}{4} \times 3\frac{3}{5} =$

28). $2\frac{1}{3} \times 4\frac{1}{2} =$

29). $2\frac{9}{10} \times 4\frac{2}{5} =$

30). $2\frac{4}{5} \times 3\frac{3}{5} =$

31). $4\frac{4}{5} \times 3\frac{1}{5} =$

32). $2\frac{1}{2} \times 3\frac{1}{2} =$

33). $3\frac{1}{2} \times 4\frac{1}{4} =$

34). $2\frac{3}{5} \times 4\frac{1}{2} =$

35). $2\frac{2}{3} \times 3\frac{4}{5} =$

36). $4\frac{2}{5} \times 2\frac{1}{2} =$

37). $4\frac{1}{4} \times 4\frac{2}{3} =$

38). $4\frac{2}{3} \times 3\frac{1}{2} =$

39). $4\frac{1}{2} \times 3\frac{1}{2} =$

40). $2\frac{1}{2} \times 3\frac{2}{5} =$

41). $4\frac{1}{10} \times 4\frac{1}{5} =$

42). $3\frac{1}{2} \times 3\frac{1}{2} =$

43). $4\frac{3}{10} \times 3\frac{1}{3} =$

44). $3\frac{3}{10} \times 4\frac{3}{4} =$

45). $4\frac{3}{4} \times 3\frac{1}{10} =$

46). $4\frac{2}{5} \times 4\frac{2}{3} =$

47). $4\frac{1}{2} \times 3\frac{3}{5} =$

48). $2\frac{1}{2} \times 2\frac{2}{3} =$

49). $3\frac{9}{10} \times 4\frac{1}{4} =$

50). $4\frac{1}{2} \times 4\frac{3}{5} =$

51). $4\frac{1}{3} \times 2\frac{3}{4} =$

52). $4\frac{1}{2} \times 2\frac{2}{5} =$

53). $4\frac{1}{2} \times 2\frac{3}{4} =$

54). $2\frac{3}{5} \times 3\frac{1}{2} =$

55). $2\frac{7}{10} \times 3\frac{1}{3} =$

56). $2\frac{1}{2} \times 4\frac{1}{10} =$

57). $2\frac{4}{5} \times 3\frac{1}{2} =$

58). $2\frac{3}{5} \times 4\frac{1}{4} =$

59). $2\frac{1}{2} \times 4\frac{2}{3} =$

60). $4\frac{3}{5} \times 4\frac{1}{3} =$

61). $2\frac{1}{2} \times 2\frac{3}{4} =$

62). $3\frac{1}{5} \times 4\frac{4}{7} =$

63). $2\frac{4}{5} \times 3\frac{1}{2} =$

64). $3\frac{2}{5} \times 3\frac{3}{10} =$

65). $2\frac{6}{7} \times 4\frac{2}{9} =$

66). $3\frac{1}{10} \times 2\frac{1}{7} =$

67). $3\frac{3}{5} \times 3\frac{2}{3} =$

68). $3\frac{1}{4} \times 2\frac{5}{6} =$

69). $2\frac{1}{2} \times 4\frac{5}{9} =$

70). $4\frac{5}{6} \times 4\frac{1}{2} =$

71). $2\frac{1}{3} \times 4\frac{1}{5} =$

72). $3\frac{2}{5} \times 3\frac{1}{4} =$

73). $3\frac{1}{5} \times 3\frac{3}{8} =$

74). $2\frac{3}{10} \times 3\frac{1}{7} =$

75). $2\frac{1}{2} \times 3\frac{9}{10} =$

76). $2\frac{2}{3} \times 2\frac{1}{3} =$

77). $3\frac{1}{3} \times 4\frac{4}{9} =$

78). $3\frac{2}{9} \times 2\frac{1}{4} =$

79). $3\frac{1}{2} \times 2\frac{4}{5} =$

80). $3\frac{1}{2} \times 2\frac{4}{5} =$

81). $2\frac{5}{9} \times 4\frac{1}{10} =$

82). $3\frac{7}{8} \times 2\frac{9}{10} =$

83). $2\frac{4}{5} \times 3\frac{3}{4} =$

84). $4\frac{5}{8} \times 2\frac{1}{2} =$

85). $2\frac{2}{9} \times 4\frac{1}{2} =$

86). $4\frac{1}{3} \times 4\frac{3}{8} =$

87). $4\frac{1}{2} \times 3\frac{2}{3} =$

88). $3\frac{5}{8} \times 4\frac{2}{5} =$

89). $3\frac{1}{10} \times 4\frac{3}{5} =$

90). $2\frac{1}{2} \times 4\frac{4}{5} =$

91). $2\frac{1}{2} \times 4\frac{7}{10} =$

92). $4\frac{7}{8} \times 4\frac{1}{5} =$

93). $2\frac{2}{9} \times 4\frac{1}{2} =$

94). $4\frac{1}{2} \times 3\frac{1}{3} =$

95). $4\frac{1}{10} \times 2\frac{3}{5} =$

96). $2\frac{1}{3} \times 3\frac{1}{5} =$

97). $2\frac{1}{2} \times 3\frac{3}{4} =$

98). $3\frac{2}{3} \times 3\frac{4}{5} =$

99). $3\frac{1}{3} \times 3\frac{1}{6} =$

100). $3\frac{1}{3} \times 2\frac{1}{3} =$

ANSWER

1). $2\frac{2}{3} \times 4\frac{2}{5} = \frac{176}{15}$

2). $4\frac{3}{4} \times 3\frac{1}{2} = \frac{133}{8}$

3). $2\frac{4}{5} \times 3\frac{2}{3} = \frac{154}{15}$

4). $3\frac{3}{5} \times 3\frac{1}{2} = \frac{126}{10}$

5). $3\frac{1}{3} \times 3\frac{1}{2} = \frac{70}{6}$

6). $4\frac{1}{3} \times 2\frac{1}{2} = \frac{65}{6}$

7). $2\frac{2}{5} \times 2\frac{1}{2} = \frac{60}{10}$

8). $3\frac{1}{2} \times 3\frac{1}{4} = \frac{91}{8}$

9). $4\frac{4}{5} \times 4\frac{1}{3} = \frac{312}{15}$

10). $4\frac{1}{2} \times 4\frac{4}{5} = \frac{216}{10}$

11). $2\frac{1}{2} \times 4\frac{1}{10} = \frac{205}{20}$

12). $2\frac{2}{3} \times 2\frac{1}{5} = \frac{88}{15}$

13). $4\frac{1}{5} \times 3\frac{1}{2} = \frac{147}{10}$

14). $3\frac{1}{2} \times 4\frac{2}{5} = \frac{154}{10}$

15). $3\frac{1}{4} \times 3\frac{1}{2} = \frac{91}{8}$

16). $4\frac{1}{4} \times 2\frac{2}{3} = \frac{136}{12}$

17). $2\frac{1}{4} \times 3\frac{3}{5} = \frac{162}{20}$

18). $2\frac{1}{3} \times 4\frac{1}{2} = \frac{63}{6}$

19). $2\frac{9}{10} \times 4\frac{2}{5} = \frac{638}{50}$

20). $2\frac{4}{5} \times 3\frac{3}{5} = \frac{252}{25}$

21). $2\frac{1}{2} \times 4\frac{1}{10} = \frac{205}{20}$

22). $2\frac{2}{3} \times 2\frac{1}{5} = \frac{88}{15}$

23). $4\frac{1}{5} \times 3\frac{1}{2} = \frac{147}{10}$

24). $3\frac{1}{2} \times 4\frac{2}{5} = \frac{154}{10}$

25). $3\frac{1}{4} \times 3\frac{1}{2} = \frac{91}{8}$

26). $4\frac{1}{4} \times 2\frac{2}{3} = \frac{136}{12}$

27). $2\frac{1}{4} \times 3\frac{3}{5} = \frac{162}{20}$

28). $2\frac{1}{3} \times 4\frac{1}{2} = \frac{63}{6}$

29). $2\frac{9}{10} \times 4\frac{2}{5} = \frac{638}{50}$

30). $2\frac{4}{5} \times 3\frac{3}{5} = \frac{252}{25}$

31). $4\frac{4}{5} \times 3\frac{1}{5} = \frac{384}{25}$

32). $2\frac{1}{2} \times 3\frac{1}{2} = \frac{35}{4}$

33). $3\frac{1}{2} \times 4\frac{1}{4} = \frac{119}{8}$

34). $2\frac{3}{5} \times 4\frac{1}{2} = \frac{117}{10}$

35). $2\frac{2}{3} \times 3\frac{4}{5} = \frac{152}{15}$

36). $4\frac{2}{5} \times 2\frac{1}{2} = \frac{110}{10}$

37). $4\frac{1}{4} \times 4\frac{2}{3} = \frac{238}{12}$

38). $4\frac{2}{3} \times 3\frac{1}{2} = \frac{98}{6}$

39). $4\frac{1}{2} \times 3\frac{1}{2} = \frac{63}{4}$

40). $2\frac{1}{2} \times 3\frac{2}{5} = \frac{85}{10}$

41). $4\frac{1}{10} \times 4\frac{1}{5} = \frac{861}{50}$

42). $3\frac{1}{2} \times 3\frac{1}{2} = \frac{49}{4}$

43). $4\frac{3}{10} \times 3\frac{1}{3} = \frac{430}{30}$

44). $3\frac{3}{10} \times 4\frac{3}{4} = \frac{627}{40}$

45). $4\frac{3}{4} \times 3\frac{1}{10} = \frac{589}{40}$

46). $4\frac{2}{5} \times 4\frac{2}{3} = \frac{308}{15}$

47). $4\frac{1}{2} \times 3\frac{3}{5} = \frac{162}{10}$

48). $2\frac{1}{2} \times 2\frac{2}{3} = \frac{40}{6}$

49). $3\frac{9}{10} \times 4\frac{1}{4} = \frac{663}{40}$

50). $4\frac{1}{2} \times 4\frac{3}{5} = \frac{207}{10}$

51). $4\frac{1}{3} \times 2\frac{3}{4} = \frac{143}{12}$

52). $4\frac{1}{2} \times 2\frac{2}{5} = \frac{108}{10}$

53). $4\frac{1}{2} \times 2\frac{3}{4} = \frac{99}{8}$

54). $2\frac{3}{5} \times 3\frac{1}{2} = \frac{91}{10}$

55). $2\frac{7}{10} \times 3\frac{1}{3} = \frac{270}{30}$

56). $2\frac{1}{2} \times 4\frac{1}{10} = \frac{205}{20}$

57). $2\frac{4}{5} \times 3\frac{1}{2} = \frac{98}{10}$

58). $2\frac{3}{5} \times 4\frac{1}{4} = \frac{221}{20}$

59). $2\frac{1}{2} \times 4\frac{2}{3} = \frac{70}{6}$

60). $4\frac{3}{5} \times 4\frac{1}{3} = \frac{299}{15}$

61). $2\frac{1}{2} \times 2\frac{3}{4} = \frac{55}{8}$

62). $3\frac{1}{5} \times 4\frac{4}{7} = \frac{512}{35}$

63). $2\frac{4}{5} \times 3\frac{1}{2} = \frac{98}{10}$

64). $3\frac{2}{5} \times 3\frac{3}{10} = \frac{561}{50}$

65). $2\frac{6}{7} \times 4\frac{2}{9} = \frac{760}{63}$

66). $3\frac{1}{10} \times 2\frac{1}{7} = \frac{465}{70}$

67). $3\frac{3}{5} \times 3\frac{2}{3} = \frac{198}{15}$

68). $3\frac{1}{4} \times 2\frac{5}{6} = \frac{221}{24}$

69). $2\frac{1}{2} \times 4\frac{5}{9} = \frac{205}{18}$

70). $4\frac{5}{6} \times 4\frac{1}{2} = \frac{261}{12}$

71). $2\frac{1}{3} \times 4\frac{1}{5} = \frac{147}{15}$

72). $3\frac{2}{5} \times 3\frac{1}{4} = \frac{221}{20}$

73). $3\frac{1}{5} \times 3\frac{3}{8} = \frac{432}{40}$

74). $2\frac{3}{10} \times 3\frac{1}{7} = \frac{506}{70}$

75). $2\frac{1}{2} \times 3\frac{9}{10} = \frac{195}{20}$

76). $2\frac{2}{3} \times 2\frac{1}{3} = \frac{56}{9}$

77). $3\frac{1}{3} \times 4\frac{4}{9} = \frac{400}{27}$

78). $3\frac{2}{9} \times 2\frac{1}{4} = \frac{261}{36}$

79). $3\frac{1}{2} \times 2\frac{4}{5} = \frac{98}{10}$

80). $3\frac{1}{2} \times 2\frac{4}{5} = \frac{98}{10}$

81). $2\frac{5}{9} \times 4\frac{1}{10} = \frac{943}{90}$

82). $3\frac{7}{8} \times 2\frac{9}{10} = \frac{899}{80}$

83). $2\frac{4}{5} \times 3\frac{3}{4} = \frac{210}{20}$

84). $4\frac{5}{8} \times 2\frac{1}{2} = \frac{185}{16}$

85). $2\frac{2}{9} \times 4\frac{1}{2} = \frac{180}{18}$

86). $4\frac{1}{3} \times 4\frac{3}{8} = \frac{455}{24}$

87). $4\frac{1}{2} \times 3\frac{2}{3} = \frac{99}{6}$

88). $3\frac{5}{8} \times 4\frac{2}{5} = \frac{638}{40}$

89). $3\frac{1}{10} \times 4\frac{3}{5} = \frac{713}{50}$

90). $2\frac{1}{2} \times 4\frac{4}{5} = \frac{120}{10}$

91). $2\frac{1}{2} \times 4\frac{7}{10} = \frac{235}{20}$

92). $4\frac{7}{8} \times 4\frac{1}{5} = \frac{819}{40}$

93). $2\frac{2}{9} \times 4\frac{1}{2} = \frac{180}{18}$

94). $4\frac{1}{2} \times 3\frac{1}{3} = \frac{90}{6}$

95). $4\frac{1}{10} \times 2\frac{3}{5} = \frac{533}{50}$

96). $2\frac{1}{3} \times 3\frac{1}{5} = \frac{112}{15}$

97). $2\frac{1}{2} \times 3\frac{3}{4} = \frac{75}{8}$

98). $3\frac{2}{3} \times 3\frac{4}{5} = \frac{209}{15}$

99). $3\frac{1}{3} \times 3\frac{1}{6} = \frac{190}{18}$

100). $3\frac{1}{3} \times 2\frac{1}{3} = \frac{70}{9}$

MULTIPLYING FRACTIONS AND WHOLE NUMBERS

1). $\frac{4}{5} \times 7 =$

2). $\frac{1}{2} \times 3 =$

3). $\frac{1}{2} \times 7 =$

4). $\frac{3}{4} \times 2 =$

5). $\frac{1}{3} \times 4 =$

6). $\frac{2}{4} \times 5 =$

7). $\frac{3}{4} \times 6 =$

8). $\frac{2}{10} \times 3 =$

9). $\frac{2}{4} \times 8 =$

10). $\frac{3}{10} \times 4 =$

11). $\frac{2}{5} \times 15 =$

12). $\frac{4}{5} \times 2 =$

13). $\frac{1}{5} \times 16 =$

14). $\frac{1}{7} \times 6 =$

15). $\frac{1}{5} \times 17 =$

16). $\frac{3}{6} \times 15 =$

17). $\frac{2}{3} \times 18 =$

18). $\frac{3}{9} \times 13 =$

19). $\frac{2}{4} \times 8 =$

20). $\frac{1}{2} \times 17 =$

21). $\dfrac{1}{3} \times 6 =$

22). $\dfrac{7}{8} \times 11 =$

23). $\dfrac{6}{7} \times 5 =$

24). $\dfrac{2}{4} \times 16 =$

25). $\dfrac{5}{9} \times 3 =$

26). $\dfrac{1}{2} \times 14 =$

27). $\dfrac{5}{9} \times 2 =$

28). $\dfrac{1}{9} \times 20 =$

29). $\dfrac{4}{5} \times 10 =$

30). $\dfrac{4}{6} \times 9 =$

31). $\dfrac{1}{3} \times 8 =$

32). $\dfrac{1}{5} \times 15 =$

33). $\dfrac{4}{5} \times 17 =$

34). $\dfrac{2}{3} \times 19 =$

35). $\dfrac{2}{4} \times 3 =$

36). $\dfrac{10}{15} \times 13 =$

37). $\dfrac{9}{12} \times 20 =$

38). $\dfrac{6}{14} \times 23 =$

39). $\dfrac{3}{7} \times 16 =$

40). $\dfrac{1}{2} \times 12 =$

41). $\dfrac{12}{14} \times 3 =$

42). $\dfrac{6}{7} \times 24 =$

43). $\dfrac{11}{15} \times 8 =$

44). $\dfrac{8}{18} \times 23 =$

45). $\dfrac{3}{9} \times 21 =$

46). $\dfrac{10}{18} \times 28 =$

47). $\dfrac{8}{20} \times 19 =$

48). $\dfrac{4}{6} \times 17 =$

49). $\dfrac{7}{8} \times 7 =$

50). $\dfrac{15}{20} \times 27 =$

ANSWER

1). $\dfrac{4}{5} \times 7 = \dfrac{28}{5}$

2). $\dfrac{1}{2} \times 3 = \dfrac{3}{2}$

3). $\dfrac{1}{2} \times 7 = \dfrac{7}{2}$

4). $\dfrac{3}{4} \times 2 = \dfrac{6}{4}$

5). $\dfrac{1}{3} \times 4 = \dfrac{4}{3}$

6). $\dfrac{2}{4} \times 5 = \dfrac{10}{4}$

7). $\dfrac{3}{4} \times 6 = \dfrac{18}{4}$

8). $\dfrac{2}{10} \times 3 = \dfrac{6}{10}$

9). $\dfrac{2}{4} \times 8 = \dfrac{16}{4}$

10). $\dfrac{3}{10} \times 4 = \dfrac{12}{10}$

11). $\dfrac{2}{5} \times 15 = \dfrac{30}{5}$

12). $\dfrac{4}{5} \times 2 = \dfrac{8}{5}$

13). $\dfrac{1}{5} \times 16 = \dfrac{16}{5}$

14). $\dfrac{1}{7} \times 6 = \dfrac{6}{7}$

15). $\dfrac{1}{5} \times 17 = \dfrac{17}{5}$

16). $\dfrac{3}{6} \times 15 = \dfrac{45}{6}$

17). $\dfrac{2}{3} \times 18 = \dfrac{36}{3}$

18). $\dfrac{3}{9} \times 13 = \dfrac{39}{9}$

19). $\dfrac{2}{4} \times 8 = \dfrac{16}{4}$

20). $\dfrac{1}{2} \times 17 = \dfrac{17}{2}$

21). $\dfrac{1}{3} \times 6 = \dfrac{6}{3}$

22). $\dfrac{7}{8} \times 11 = \dfrac{77}{8}$

23). $\dfrac{6}{7} \times 5 = \dfrac{30}{7}$

24). $\dfrac{2}{4} \times 16 = \dfrac{32}{4}$

25). $\dfrac{5}{9} \times 3 = \dfrac{15}{9}$

26). $\dfrac{1}{2} \times 14 = \dfrac{14}{2}$

27). $\dfrac{5}{9} \times 2 = \dfrac{10}{9}$

28). $\dfrac{1}{9} \times 20 = \dfrac{20}{9}$

29). $\dfrac{4}{5} \times 10 = \dfrac{40}{5}$

30). $\dfrac{4}{6} \times 9 = \dfrac{36}{6}$

31). $\dfrac{1}{3} \times 8 = \dfrac{8}{3}$

32). $\dfrac{1}{5} \times 15 = \dfrac{15}{5}$

33). $\dfrac{4}{5} \times 17 = \dfrac{68}{5}$

34). $\dfrac{2}{3} \times 19 = \dfrac{38}{3}$

35). $\dfrac{2}{4} \times 3 = \dfrac{6}{4}$

36). $\dfrac{10}{15} \times 13 = \dfrac{130}{15}$

37). $\dfrac{9}{12} \times 20 = \dfrac{180}{12}$

38). $\dfrac{6}{14} \times 23 = \dfrac{138}{14}$

39). $\dfrac{3}{7} \times 16 = \dfrac{48}{7}$

40). $\dfrac{1}{2} \times 12 = \dfrac{12}{2}$

41). $\dfrac{12}{14} \times 3 = \dfrac{36}{14}$

42). $\dfrac{6}{7} \times 24 = \dfrac{144}{7}$

43). $\dfrac{11}{15} \times 8 = \dfrac{88}{15}$

44). $\dfrac{8}{18} \times 23 = \dfrac{184}{18}$

45). $\dfrac{3}{9} \times 21 = \dfrac{63}{9}$

46). $\dfrac{10}{18} \times 28 = \dfrac{280}{18}$

47). $\dfrac{8}{20} \times 19 = \dfrac{152}{20}$

48). $\dfrac{4}{6} \times 17 = \dfrac{68}{6}$

49). $\dfrac{7}{8} \times 7 = \dfrac{49}{8}$

50). $\dfrac{15}{20} \times 27 = \dfrac{405}{20}$

DIVISION OF FRACTIONS

1). $\dfrac{3}{4} \div \dfrac{1}{5} =$

2). $\dfrac{3}{10} \div \dfrac{2}{5} =$

3). $\dfrac{4}{10} \div \dfrac{1}{4} =$

4). $\dfrac{2}{5} \div \dfrac{2}{3} =$

5). $\dfrac{1}{2} \div \dfrac{8}{10} =$

6). $\dfrac{1}{2} \div \dfrac{2}{5} =$

7). $\dfrac{3}{5} \div \dfrac{2}{4} =$

8). $\dfrac{2}{5} \div \dfrac{1}{2} =$

9). $\dfrac{1}{4} \div \dfrac{1}{5} =$

10). $\dfrac{2}{5} \div \dfrac{1}{3} =$

11). $\dfrac{3}{4} \div \dfrac{1}{3} =$

12). $\dfrac{2}{3} \div \dfrac{1}{4} =$

13). $\dfrac{1}{10} \div \dfrac{1}{3} =$

14). $\dfrac{1}{2} \div \dfrac{2}{3} =$

15). $\dfrac{8}{10} \div \dfrac{1}{5} =$

16). $\dfrac{1}{2} \div \dfrac{2}{10} =$

17). $\dfrac{1}{4} \div \dfrac{2}{10} =$

18). $\dfrac{1}{3} \div \dfrac{8}{10} =$

19). $\dfrac{2}{10} \div \dfrac{1}{5} =$

20). $\dfrac{1}{2} \div \dfrac{3}{5} =$

21). $\dfrac{2}{3} \div \dfrac{6}{7} =$

22). $\dfrac{5}{9} \div \dfrac{1}{5} =$

23). $\dfrac{6}{7} \div \dfrac{5}{10} =$

24). $\dfrac{2}{4} \div \dfrac{1}{7} =$

25). $\dfrac{3}{7} \div \dfrac{2}{4} =$

26). $\dfrac{2}{3} \div \dfrac{4}{10} =$

27). $\dfrac{5}{6} \div \dfrac{1}{3} =$

28). $\dfrac{3}{9} \div \dfrac{2}{6} =$

29). $\dfrac{3}{9} \div \dfrac{1}{4} =$

30). $\dfrac{4}{6} \div \dfrac{2}{3} =$

31). $\dfrac{7}{8} \div \dfrac{6}{7} =$

32). $\dfrac{1}{3} \div \dfrac{1}{2} =$

33). $\dfrac{2}{8} \div \dfrac{2}{4} =$

34). $\dfrac{4}{9} \div \dfrac{2}{8} =$

35). $\dfrac{1}{2} \div \dfrac{3}{4} =$

36). $\dfrac{1}{2} \div \dfrac{1}{7} =$

37). $\dfrac{1}{5} \div \dfrac{2}{7} =$

38). $\dfrac{5}{10} \div \dfrac{4}{8} =$

39). $\dfrac{3}{4} \div \dfrac{2}{5} =$

40). $\dfrac{1}{2} \div \dfrac{3}{10} =$

41). $\dfrac{4}{6} \div \dfrac{7}{9} =$

42). $\dfrac{4}{5} \div \dfrac{1}{10} =$

43). $\dfrac{2}{5} \div \dfrac{8}{10} =$

44). $\dfrac{3}{4} \div \dfrac{2}{5} =$

45). $\dfrac{2}{5} \div \dfrac{2}{3} =$

46). $\dfrac{2}{5} \div \dfrac{1}{3} =$

47). $\dfrac{5}{8} \div \dfrac{5}{6} =$

48). $\dfrac{2}{3} \div \dfrac{3}{4} =$

49). $\dfrac{7}{9} \div \dfrac{1}{2} =$

50). $\dfrac{5}{9} \div \dfrac{1}{4} =$

51). $\dfrac{1}{5} \div \dfrac{4}{7} =$

52). $\dfrac{6}{8} \div \dfrac{7}{9} =$

53). $\dfrac{1}{5} \div \dfrac{4}{6} =$

54). $\dfrac{6}{10} \div \dfrac{4}{6} =$

55). $\dfrac{2}{9} \div \dfrac{3}{10} =$

56). $\dfrac{1}{3} \div \dfrac{3}{4} =$

57). $\dfrac{6}{8} \div \dfrac{3}{9} =$

58). $\dfrac{1}{5} \div \dfrac{1}{10} =$

59). $\dfrac{3}{4} \div \dfrac{2}{3} =$

60). $\dfrac{3}{4} \div \dfrac{1}{2} =$

61). $\dfrac{1}{5} \div \dfrac{4}{7} =$

62). $\dfrac{6}{8} \div \dfrac{7}{9} =$

63). $\dfrac{1}{5} \div \dfrac{4}{6} =$

64). $\dfrac{6}{10} \div \dfrac{4}{6} =$

65). $\dfrac{2}{9} \div \dfrac{3}{10} =$

66). $\dfrac{1}{3} \div \dfrac{3}{4} =$

67). $\dfrac{6}{8} \div \dfrac{3}{9} =$

68). $\dfrac{1}{5} \div \dfrac{1}{10} =$

69). $\dfrac{3}{4} \div \dfrac{2}{3} =$

70). $\dfrac{3}{4} \div \dfrac{1}{2} =$

71). $\dfrac{2}{5} \div \dfrac{8}{9} =$

72). $\dfrac{1}{2} \div \dfrac{5}{8} =$

73). $\dfrac{1}{2} \div \dfrac{1}{4} =$

74). $\dfrac{13}{18} \div \dfrac{1}{9} =$

75). $\dfrac{5}{10} \div \dfrac{3}{7} =$

76). $\dfrac{10}{18} \div \dfrac{10}{14} =$

77). $\dfrac{2}{5} \div \dfrac{6}{12} =$

78). $\dfrac{7}{8} \div \dfrac{8}{16} =$

79). $\dfrac{3}{5} \div \dfrac{10}{20} =$

80). $\dfrac{6}{7} \div \dfrac{2}{5} =$

81). $\dfrac{6}{10} \div \dfrac{3}{9} =$

82). $\dfrac{5}{14} \div \dfrac{1}{10} =$

83). $\dfrac{13}{20} \div \dfrac{4}{7} =$

84). $\dfrac{6}{7} \div \dfrac{12}{20} =$

85). $\dfrac{1}{2} \div \dfrac{4}{14} =$

86). $\dfrac{1}{4} \div \dfrac{8}{9} =$

87). $\dfrac{6}{16} \div \dfrac{12}{14} =$

88). $\dfrac{2}{6} \div \dfrac{5}{15} =$

89). $\dfrac{1}{2} \div \dfrac{3}{20} =$

90). $\dfrac{14}{16} \div \dfrac{5}{6} =$

91). $\dfrac{1}{2} \div \dfrac{14}{16} =$

92). $\dfrac{5}{15} \div \dfrac{1}{18} =$

93). $\dfrac{4}{5} \div \dfrac{5}{12} =$

94). $\dfrac{2}{4} \div \dfrac{5}{7} =$

95). $\dfrac{3}{15} \div \dfrac{3}{6} =$

96). $\dfrac{5}{8} \div \dfrac{3}{5} =$

97). $\dfrac{3}{15} \div \dfrac{2}{5} =$

98). $\dfrac{3}{4} \div \dfrac{18}{20} =$

99). $\dfrac{1}{2} \div \dfrac{6}{16} =$

100). $\dfrac{11}{18} \div \dfrac{6}{12} =$

ANSWER

1). $\dfrac{3}{4} \div \dfrac{1}{5} = \dfrac{15}{4}$

2). $\dfrac{3}{10} \div \dfrac{2}{5} = \dfrac{15}{20}$

3). $\dfrac{4}{10} \div \dfrac{1}{4} = \dfrac{16}{10}$

4). $\dfrac{2}{5} \div \dfrac{2}{3} = \dfrac{6}{10}$

5). $\dfrac{1}{2} \div \dfrac{8}{10} = \dfrac{10}{16}$

6). $\dfrac{1}{2} \div \dfrac{2}{5} = \dfrac{5}{4}$

7). $\dfrac{3}{5} \div \dfrac{2}{4} = \dfrac{12}{10}$

8). $\dfrac{2}{5} \div \dfrac{1}{2} = \dfrac{4}{5}$

9). $\dfrac{1}{4} \div \dfrac{1}{5} = \dfrac{5}{4}$

10). $\dfrac{2}{5} \div \dfrac{1}{3} = \dfrac{6}{5}$

11). $\dfrac{3}{4} \div \dfrac{1}{3} = \dfrac{9}{4}$

12). $\dfrac{2}{3} \div \dfrac{1}{4} = \dfrac{8}{3}$

13). $\dfrac{1}{10} \div \dfrac{1}{3} = \dfrac{3}{10}$

14). $\dfrac{1}{2} \div \dfrac{2}{3} = \dfrac{3}{4}$

15). $\dfrac{8}{10} \div \dfrac{1}{5} = \dfrac{40}{10}$

16). $\dfrac{1}{2} \div \dfrac{2}{10} = \dfrac{10}{4}$

17). $\dfrac{1}{4} \div \dfrac{2}{10} = \dfrac{10}{8}$

18). $\dfrac{1}{3} \div \dfrac{8}{10} = \dfrac{10}{24}$

19). $\dfrac{2}{10} \div \dfrac{1}{5} = \dfrac{10}{10}$

20). $\dfrac{1}{2} \div \dfrac{3}{5} = \dfrac{5}{6}$

21). $\dfrac{2}{3} \div \dfrac{6}{7} = \dfrac{14}{18}$

22). $\dfrac{5}{9} \div \dfrac{1}{5} = \dfrac{25}{9}$

23). $\dfrac{6}{7} \div \dfrac{5}{10} = \dfrac{60}{35}$

24). $\dfrac{2}{4} \div \dfrac{1}{7} = \dfrac{14}{4}$

25). $\dfrac{3}{7} \div \dfrac{2}{4} = \dfrac{12}{14}$

26). $\dfrac{2}{3} \div \dfrac{4}{10} = \dfrac{20}{12}$

27). $\dfrac{5}{6} \div \dfrac{1}{3} = \dfrac{15}{6}$

28). $\dfrac{3}{9} \div \dfrac{2}{6} = \dfrac{18}{18}$

29). $\dfrac{3}{9} \div \dfrac{1}{4} = \dfrac{12}{9}$

30). $\dfrac{4}{6} \div \dfrac{2}{3} = \dfrac{12}{12}$

31). $\dfrac{7}{8} \div \dfrac{6}{7} = \dfrac{49}{48}$

32). $\dfrac{1}{3} \div \dfrac{1}{2} = \dfrac{2}{3}$

33). $\dfrac{2}{8} \div \dfrac{2}{4} = \dfrac{8}{16}$

34). $\dfrac{4}{9} \div \dfrac{2}{8} = \dfrac{32}{18}$

35). $\dfrac{1}{2} \div \dfrac{3}{4} = \dfrac{4}{6}$

36). $\dfrac{1}{2} \div \dfrac{1}{7} = \dfrac{7}{2}$

37). $\dfrac{1}{5} \div \dfrac{2}{7} = \dfrac{7}{10}$

38). $\dfrac{5}{10} \div \dfrac{4}{8} = \dfrac{40}{40}$

39). $\dfrac{3}{4} \div \dfrac{2}{5} = \dfrac{15}{8}$

40). $\dfrac{1}{2} \div \dfrac{3}{10} = \dfrac{10}{6}$

41). $\dfrac{4}{6} \div \dfrac{7}{9} = \dfrac{36}{42}$

42). $\dfrac{4}{5} \div \dfrac{1}{10} = \dfrac{40}{5}$

43). $\dfrac{2}{5} \div \dfrac{8}{10} = \dfrac{20}{40}$

44). $\dfrac{3}{4} \div \dfrac{2}{5} = \dfrac{15}{8}$

45). $\dfrac{2}{5} \div \dfrac{2}{3} = \dfrac{6}{10}$

46). $\dfrac{2}{5} \div \dfrac{1}{3} = \dfrac{6}{5}$

47). $\dfrac{5}{8} \div \dfrac{5}{6} = \dfrac{30}{40}$

48). $\dfrac{2}{3} \div \dfrac{3}{4} = \dfrac{8}{9}$

49). $\dfrac{7}{9} \div \dfrac{1}{2} = \dfrac{14}{9}$

50). $\dfrac{5}{9} \div \dfrac{1}{4} = \dfrac{20}{9}$

51). $\dfrac{1}{5} \div \dfrac{4}{7} = \dfrac{7}{20}$

52). $\dfrac{6}{8} \div \dfrac{7}{9} = \dfrac{54}{56}$

53). $\dfrac{1}{5} \div \dfrac{4}{6} = \dfrac{6}{20}$

54). $\dfrac{6}{10} \div \dfrac{4}{6} = \dfrac{36}{40}$

55). $\dfrac{2}{9} \div \dfrac{3}{10} = \dfrac{20}{27}$

56). $\dfrac{1}{3} \div \dfrac{3}{4} = \dfrac{4}{9}$

57). $\dfrac{6}{8} \div \dfrac{3}{9} = \dfrac{54}{24}$

58). $\dfrac{1}{5} \div \dfrac{1}{10} = \dfrac{10}{5}$

59). $\dfrac{3}{4} \div \dfrac{2}{3} = \dfrac{9}{8}$

60). $\dfrac{3}{4} \div \dfrac{1}{2} = \dfrac{6}{4}$

61). $\dfrac{1}{5} \div \dfrac{4}{7} = \dfrac{7}{20}$

62). $\dfrac{6}{8} \div \dfrac{7}{9} = \dfrac{54}{56}$

63). $\dfrac{1}{5} \div \dfrac{4}{6} = \dfrac{6}{20}$

64). $\dfrac{6}{10} \div \dfrac{4}{6} = \dfrac{36}{40}$

65). $\dfrac{2}{9} \div \dfrac{3}{10} = \dfrac{20}{27}$

66). $\dfrac{1}{3} \div \dfrac{3}{4} = \dfrac{4}{9}$

67). $\dfrac{6}{8} \div \dfrac{3}{9} = \dfrac{54}{24}$

68). $\dfrac{1}{5} \div \dfrac{1}{10} = \dfrac{10}{5}$

69). $\dfrac{3}{4} \div \dfrac{2}{3} = \dfrac{9}{8}$

70). $\dfrac{3}{4} \div \dfrac{1}{2} = \dfrac{6}{4}$

71). $\dfrac{2}{5} \div \dfrac{8}{9} = \dfrac{18}{40}$

72). $\dfrac{1}{2} \div \dfrac{5}{8} = \dfrac{8}{10}$

73). $\dfrac{1}{2} \div \dfrac{1}{4} = \dfrac{4}{2}$

74). $\dfrac{13}{18} \div \dfrac{1}{9} = \dfrac{117}{18}$

75). $\dfrac{5}{10} \div \dfrac{3}{7} = \dfrac{35}{30}$

76). $\dfrac{10}{18} \div \dfrac{10}{14} = \dfrac{140}{180}$

77). $\dfrac{2}{5} \div \dfrac{6}{12} = \dfrac{24}{30}$

78). $\dfrac{7}{8} \div \dfrac{8}{16} = \dfrac{112}{64}$

79). $\dfrac{3}{5} \div \dfrac{10}{20} = \dfrac{60}{50}$

80). $\dfrac{6}{7} \div \dfrac{2}{5} = \dfrac{30}{14}$

81). $\dfrac{6}{10} \div \dfrac{3}{9} = \dfrac{54}{30}$

82). $\dfrac{5}{14} \div \dfrac{1}{10} = \dfrac{50}{14}$

83). $\dfrac{13}{20} \div \dfrac{4}{7} = \dfrac{91}{80}$

84). $\dfrac{6}{7} \div \dfrac{12}{20} = \dfrac{120}{84}$

85). $\dfrac{1}{2} \div \dfrac{4}{14} = \dfrac{14}{8}$

86). $\dfrac{1}{4} \div \dfrac{8}{9} = \dfrac{9}{32}$

87). $\dfrac{6}{16} \div \dfrac{12}{14} = \dfrac{84}{192}$

88). $\dfrac{2}{6} \div \dfrac{5}{15} = \dfrac{30}{30}$

89). $\dfrac{1}{2} \div \dfrac{3}{20} = \dfrac{20}{6}$

90). $\dfrac{14}{16} \div \dfrac{5}{6} = \dfrac{84}{80}$

91). $\dfrac{1}{2} \div \dfrac{14}{16} = \dfrac{16}{28}$

92). $\dfrac{5}{15} \div \dfrac{1}{18} = \dfrac{90}{15}$

93). $\dfrac{4}{5} \div \dfrac{5}{12} = \dfrac{48}{25}$

94). $\dfrac{2}{4} \div \dfrac{5}{7} = \dfrac{14}{20}$

95). $\dfrac{3}{15} \div \dfrac{3}{6} = \dfrac{18}{45}$

96). $\dfrac{5}{8} \div \dfrac{3}{5} = \dfrac{25}{24}$

97). $\dfrac{3}{15} \div \dfrac{2}{5} = \dfrac{15}{30}$

98). $\dfrac{3}{4} \div \dfrac{18}{20} = \dfrac{60}{72}$

99). $\dfrac{1}{2} \div \dfrac{6}{16} = \dfrac{16}{12}$

100). $\dfrac{11}{18} \div \dfrac{6}{12} = \dfrac{132}{108}$

DIVISION MIXED FRACTIONS

1). $3\frac{2}{5} \div 2\frac{1}{2} =$

2). $3\frac{2}{3} \div 2\frac{1}{4} =$

3). $2\frac{1}{4} \div 4\frac{2}{5} =$

4). $3\frac{1}{5} \div 4\frac{1}{2} =$

5). $2\frac{1}{2} \div 2\frac{1}{2} =$

6). $2\frac{1}{5} \div 3\frac{1}{10} =$

7). $2\frac{3}{4} \div 4\frac{1}{2} =$

8). $3\frac{3}{5} \div 3\frac{1}{3} =$

9). $2\frac{2}{5} \div 2\frac{2}{3} =$

10). $2\frac{1}{2} \div 3\frac{4}{5} =$

11). $4\frac{3}{4} \div 4\frac{2}{3} =$

12). $4\frac{2}{5} \div 2\frac{1}{2} =$

13). $3\frac{2}{5} \div 3\frac{1}{2} =$

14). $3\frac{1}{3} \div 4\frac{3}{4} =$

15). $3\frac{1}{10} \div 4\frac{1}{5} =$

16). $2\frac{1}{10} \div 2\frac{3}{4} =$

17). $2\frac{3}{5} \div 4\frac{1}{3} =$

18). $3\frac{1}{2} \div 2\frac{1}{5} =$

19). $4\frac{4}{5} \div 2\frac{1}{2} =$

20). $4\frac{2}{5} \div 3\frac{2}{5} =$

21). $2\frac{1}{2} \div 3\frac{2}{3} =$

22). $4\frac{3}{5} \div 3\frac{1}{2} =$

23). $2\frac{1}{3} \div 3\frac{1}{2} =$

24). $2\frac{4}{5} \div 2\frac{3}{4} =$

25). $3\frac{1}{3} \div 4\frac{3}{5} =$

26). $2\frac{4}{5} \div 2\frac{1}{3} =$

27). $2\frac{1}{4} \div 4\frac{2}{5} =$

28). $2\frac{2}{5} \div 4\frac{3}{5} =$

29). $3\frac{1}{2} \div 4\frac{3}{4} =$

30). $3\frac{2}{3} \div 3\frac{1}{5} =$

31). $4\frac{1}{2} \div 2\frac{1}{4} =$

32). $2\frac{7}{10} \div 3\frac{2}{5} =$

33). $4\frac{3}{5} \div 4\frac{1}{2} =$

34). $2\frac{3}{5} \div 3\frac{2}{3} =$

35). $3\frac{1}{2} \div 2\frac{2}{5} =$

36). $4\frac{1}{2} \div 3\frac{3}{4} =$

37). $2\frac{1}{2} \div 3\frac{3}{5} =$

38). $4\frac{1}{3} \div 3\frac{1}{5} =$

39). $3\frac{1}{2} \div 4\frac{1}{4} =$

40). $4\frac{1}{2} \div 2\frac{1}{4} =$

41). $3\frac{1}{3} \div 4\frac{4}{5} =$

42). $2\frac{1}{10} \div 2\frac{1}{4} =$

43). $2\frac{1}{5} \div 3\frac{1}{2} =$

44). $4\frac{2}{3} \div 2\frac{1}{2} =$

45). $3\frac{1}{10} \div 3\frac{3}{5} =$

46). $2\frac{3}{5} \div 2\frac{1}{5} =$

47). $3\frac{2}{3} \div 3\frac{1}{2} =$

48). $4\frac{1}{3} \div 3\frac{1}{2} =$

49). $2\frac{1}{5} \div 4\frac{1}{3} =$

50). $4\frac{9}{10} \div 2\frac{1}{4} =$

51). $4\frac{1}{5} \div 4\frac{1}{3} =$

52). $3\frac{1}{3} \div 2\frac{3}{5} =$

53). $4\frac{2}{3} \div 4\frac{1}{2} =$

54). $2\frac{1}{3} \div 2\frac{1}{2} =$

55). $2\frac{2}{5} \div 2\frac{4}{5} =$

56). $3\frac{4}{5} \div 2\frac{1}{2} =$

57). $3\frac{3}{5} \div 3\frac{1}{5} =$

58). $4\frac{1}{2} \div 4\frac{3}{5} =$

59). $2\frac{1}{2} \div 4\frac{1}{2} =$

60). $4\frac{1}{2} \div 3\frac{3}{4} =$

61). $3\frac{4}{9} \div 3\frac{1}{2} =$

62). $3\frac{1}{2} \div 3\frac{2}{3} =$

63). $4\frac{1}{2} \div 2\frac{2}{5} =$

64). $2\frac{1}{2} \div 2\frac{1}{5} =$

65). $4\frac{2}{5} \div 2\frac{1}{2} =$

66). $2\frac{2}{3} \div 2\frac{3}{10} =$

67). $2\frac{5}{6} \div 3\frac{1}{7} =$

68). $4\frac{7}{9} \div 3\frac{2}{3} =$

69). $2\frac{2}{3} \div 3\frac{3}{7} =$

70). $4\frac{3}{8} \div 2\frac{1}{6} =$

71). $2\frac{1}{2} \div 2\frac{1}{2} =$

72). $2\frac{1}{2} \div 4\frac{4}{5} =$

73). $2\frac{1}{4} \div 2\frac{1}{2} =$

74). $2\frac{3}{7} \div 4\frac{1}{3} =$

75). $4\frac{1}{9} \div 4\frac{3}{4} =$

76). $2\frac{5}{7} \div 3\frac{1}{2} =$

77). $3\frac{2}{7} \div 4\frac{1}{4} =$

78). $2\frac{2}{3} \div 3\frac{1}{3} =$

79). $4\frac{1}{2} \div 3\frac{1}{2} =$

80). $2\frac{5}{9} \div 3\frac{1}{4} =$

81). $4\frac{2}{5} \div 4\frac{9}{10} =$

82). $3\frac{3}{4} \div 3\frac{2}{5} =$

83). $4\frac{3}{4} \div 4\frac{1}{5} =$

84). $4\frac{1}{2} \div 2\frac{1}{10} =$

85). $4\frac{4}{7} \div 2\frac{1}{2} =$

86). $3\frac{1}{6} \div 4\frac{3}{5} =$

87). $4\frac{4}{7} \div 4\frac{1}{3} =$

88). $3\frac{1}{3} \div 3\frac{2}{5} =$

89). $3\frac{1}{2} \div 3\frac{5}{8} =$

90). $4\frac{1}{3} \div 4\frac{6}{7} =$

91). $4\frac{3}{4} \div 4\frac{1}{9} =$

92). $3\frac{2}{3} \div 3\frac{3}{4} =$

93). $2\frac{4}{5} \div 4\frac{1}{3} =$

94). $4\frac{9}{10} \div 2\frac{1}{4} =$

95). $4\frac{1}{2} \div 2\frac{2}{3} =$

96). $2\frac{2}{3} \div 2\frac{3}{10} =$

97). $2\frac{2}{3} \div 4\frac{5}{6} =$

98). $2\frac{5}{7} \div 2\frac{1}{6} =$

99). $4\frac{1}{3} \div 3\frac{2}{3} =$

100). $4\frac{1}{7} \div 3\frac{1}{6} =$

ANSWER

1). $3\frac{2}{5} \div 2\frac{1}{2} = \frac{34}{25}$

2). $3\frac{2}{3} \div 2\frac{1}{4} = \frac{44}{27}$

3). $2\frac{1}{4} \div 4\frac{2}{5} = \frac{45}{88}$

4). $3\frac{1}{5} \div 4\frac{1}{2} = \frac{32}{45}$

5). $2\frac{1}{2} \div 2\frac{1}{2} = \frac{10}{10}$

6). $2\frac{1}{5} \div 3\frac{1}{10} = \frac{110}{155}$

7). $2\frac{3}{4} \div 4\frac{1}{2} = \frac{22}{36}$

8). $3\frac{3}{5} \div 3\frac{1}{3} = \frac{54}{50}$

9). $2\frac{2}{5} \div 2\frac{2}{3} = \frac{36}{40}$

10). $2\frac{1}{2} \div 3\frac{4}{5} = \frac{25}{38}$

11). $4\frac{3}{4} \div 4\frac{2}{3} = \frac{57}{56}$

12). $4\frac{2}{5} \div 2\frac{1}{2} = \frac{44}{25}$

13). $3\frac{2}{5} \div 3\frac{1}{2} = \frac{34}{35}$

14). $3\frac{1}{3} \div 4\frac{3}{4} = \frac{40}{57}$

15). $3\frac{1}{10} \div 4\frac{1}{5} = \frac{155}{210}$

16). $2\frac{1}{10} \div 2\frac{3}{4} = \frac{84}{110}$

17). $2\frac{3}{5} \div 4\frac{1}{3} = \frac{39}{65}$

18). $3\frac{1}{2} \div 2\frac{1}{5} = \frac{35}{22}$

19). $4\frac{4}{5} \div 2\frac{1}{2} = \frac{48}{25}$

20). $4\frac{2}{5} \div 3\frac{2}{5} = \frac{110}{85}$

21). $2\frac{1}{2} \div 3\frac{2}{3} = \frac{15}{22}$

22). $4\frac{3}{5} \div 3\frac{1}{2} = \frac{46}{35}$

23). $2\frac{1}{3} \div 3\frac{1}{2} = \frac{14}{21}$

24). $2\frac{4}{5} \div 2\frac{3}{4} = \frac{56}{55}$

25). $3\frac{1}{3} \div 4\frac{3}{5} = \frac{50}{69}$

26). $2\frac{4}{5} \div 2\frac{1}{3} = \frac{42}{35}$

27). $2\frac{1}{4} \div 4\frac{2}{5} = \frac{45}{88}$

28). $2\frac{2}{5} \div 4\frac{3}{5} = \frac{60}{115}$

29). $3\frac{1}{2} \div 4\frac{3}{4} = \frac{28}{38}$

30). $3\frac{2}{3} \div 3\frac{1}{5} = \frac{55}{48}$

31). $4\frac{1}{2} \div 2\frac{1}{4} = \frac{36}{18}$

32). $2\frac{7}{10} \div 3\frac{2}{5} = \frac{135}{170}$

33). $4\frac{3}{5} \div 4\frac{1}{2} = \frac{46}{45}$

34). $2\frac{3}{5} \div 3\frac{2}{3} = \frac{39}{55}$

35). $3\frac{1}{2} \div 2\frac{2}{5} = \frac{35}{24}$

36). $4\frac{1}{2} \div 3\frac{3}{4} = \frac{36}{30}$

37). $2\frac{1}{2} \div 3\frac{3}{5} = \frac{25}{36}$

38). $4\frac{1}{3} \div 3\frac{1}{5} = \frac{65}{48}$

39). $3\frac{1}{2} \div 4\frac{1}{4} = \frac{28}{34}$

40). $4\frac{1}{2} \div 2\frac{1}{4} = \frac{36}{18}$

41). $3\frac{1}{3} \div 4\frac{4}{5} = \frac{50}{72}$

42). $2\frac{1}{10} \div 2\frac{1}{4} = \frac{84}{90}$

43). $2\frac{1}{5} \div 3\frac{1}{2} = \frac{22}{35}$

44). $4\frac{2}{3} \div 2\frac{1}{2} = \frac{28}{15}$

45). $3\frac{1}{10} \div 3\frac{3}{5} = \frac{155}{180}$

46). $2\frac{3}{5} \div 2\frac{1}{5} = \frac{65}{55}$

47). $3\frac{2}{3} \div 3\frac{1}{2} = \frac{22}{21}$

48). $4\frac{1}{3} \div 3\frac{1}{2} = \frac{26}{21}$

49). $2\frac{1}{5} \div 4\frac{1}{3} = \frac{33}{65}$

50). $4\frac{9}{10} \div 2\frac{1}{4} = \frac{196}{90}$

51). $4\frac{1}{5} \div 4\frac{1}{3} = \frac{63}{65}$

52). $3\frac{1}{3} \div 2\frac{3}{5} = \frac{50}{39}$

53). $4\frac{2}{3} \div 4\frac{1}{2} = \frac{28}{27}$

54). $2\frac{1}{3} \div 2\frac{1}{2} = \frac{14}{15}$

55). $2\frac{2}{5} \div 2\frac{4}{5} = \frac{60}{70}$

56). $3\frac{4}{5} \div 2\frac{1}{2} = \frac{38}{25}$

57). $3\frac{3}{5} \div 3\frac{1}{5} = \frac{90}{80}$

58). $4\frac{1}{2} \div 4\frac{3}{5} = \frac{45}{46}$

59). $2\frac{1}{2} \div 4\frac{1}{2} = \frac{10}{18}$

60). $4\frac{1}{2} \div 3\frac{3}{4} = \frac{36}{30}$

61). $3\frac{4}{9} \div 3\frac{1}{2} = \frac{62}{63}$

62). $3\frac{1}{2} \div 3\frac{2}{3} = \frac{21}{22}$

63). $4\frac{1}{2} \div 2\frac{2}{5} = \frac{45}{24}$

64). $2\frac{1}{2} \div 2\frac{1}{5} = \frac{25}{22}$

65). $4\frac{2}{5} \div 2\frac{1}{2} = \frac{44}{25}$

66). $2\frac{2}{3} \div 2\frac{3}{10} = \frac{80}{69}$

67). $2\frac{5}{6} \div 3\frac{1}{7} = \frac{119}{132}$

68). $4\frac{7}{9} \div 3\frac{2}{3} = \frac{129}{99}$

69). $2\frac{2}{3} \div 3\frac{3}{7} = \frac{56}{72}$

70). $4\frac{3}{8} \div 2\frac{1}{6} = \frac{210}{104}$

71). $2\frac{1}{2} \div 2\frac{1}{2} = \frac{10}{10}$

72). $2\frac{1}{2} \div 4\frac{4}{5} = \frac{25}{48}$

73). $2\frac{1}{4} \div 2\frac{1}{2} = \frac{18}{20}$

74). $2\frac{3}{7} \div 4\frac{1}{3} = \frac{51}{91}$

75). $4\frac{1}{9} \div 4\frac{3}{4} = \frac{148}{171}$

76). $2\frac{5}{7} \div 3\frac{1}{2} = \frac{38}{49}$

77). $3\frac{2}{7} \div 4\frac{1}{4} = \frac{92}{119}$

78). $2\frac{2}{3} \div 3\frac{1}{3} = \frac{24}{30}$

79). $4\frac{1}{2} \div 3\frac{1}{2} = \frac{18}{14}$

80). $2\frac{5}{9} \div 3\frac{1}{4} = \frac{92}{117}$

81). $4\frac{2}{5} \div 4\frac{9}{10} = \frac{220}{245}$

82). $3\frac{3}{4} \div 3\frac{2}{5} = \frac{75}{68}$

83). $4\frac{3}{4} \div 4\frac{1}{5} = \frac{95}{84}$

84). $4\frac{1}{2} \div 2\frac{1}{10} = \frac{90}{42}$

85). $4\frac{4}{7} \div 2\frac{1}{2} = \frac{64}{35}$

86). $3\frac{1}{6} \div 4\frac{3}{5} = \frac{95}{138}$

87). $4\frac{4}{7} \div 4\frac{1}{3} = \frac{96}{91}$

88). $3\frac{1}{3} \div 3\frac{2}{5} = \frac{50}{51}$

89). $3\frac{1}{2} \div 3\frac{5}{8} = \frac{56}{58}$

90). $4\frac{1}{3} \div 4\frac{6}{7} = \frac{91}{102}$

91). $4\frac{3}{4} \div 4\frac{1}{9} = \frac{171}{148}$

92). $3\frac{2}{3} \div 3\frac{3}{4} = \frac{44}{45}$

93). $2\frac{4}{5} \div 4\frac{1}{3} = \frac{42}{65}$

94). $4\frac{9}{10} \div 2\frac{1}{4} = \frac{196}{90}$

95). $4\frac{1}{2} \div 2\frac{2}{3} = \frac{27}{16}$

96). $2\frac{2}{3} \div 2\frac{3}{10} = \frac{80}{69}$

97). $2\frac{2}{3} \div 4\frac{5}{6} = \frac{48}{87}$

98). $2\frac{5}{7} \div 2\frac{1}{6} = \frac{114}{91}$

99). $4\frac{1}{3} \div 3\frac{2}{3} = \frac{39}{33}$

100). $4\frac{1}{7} \div 3\frac{1}{6} = \frac{174}{133}$

Dividing Fractions and Whole Numbers

1). $\frac{1}{2} \div 6 =$

2). $\frac{1}{4} \div 5 =$

3). $\frac{2}{3} \div 4 =$

4). $\frac{2}{5} \div 2 =$

5). $\frac{1}{5} \div 10 =$

6). $\frac{5}{10} \div 7 =$

7). $\frac{2}{4} \div 7 =$

8). $\frac{2}{4} \div 5 =$

9). $7 \div \frac{1}{2} =$

10). $7 \div \frac{2}{4} =$

11). $6 \div \frac{1}{4} =$

12). $5 \div \frac{2}{4} =$

13). $\frac{3}{5} \div 6 =$

14). $\frac{1}{2} \div 6 =$

15). $\frac{2}{4} \div 7 =$

16). $\frac{4}{10} \div 7 =$

17). $2 \div \frac{1}{5} =$

18). $5 \div \frac{1}{2} =$

19). $10 \div \frac{2}{10} =$

20). $4 \div \frac{1}{2} =$

21). $11 \div \frac{5}{8} =$

22). $2 \div \frac{3}{8} =$

23). $\frac{2}{9} \div 17 =$

24). $10 \div \frac{4}{6} =$

25). $\frac{4}{7} \div 2 =$

26). $10 \div \frac{1}{8} =$

27). $\frac{1}{6} \div 13 =$

28). $\frac{1}{4} \div 2 =$

29). $\frac{2}{5} \div 13 =$

30). $13 \div \frac{3}{6} =$

31). $\frac{1}{3} \div 20 =$

32). $2 \div \frac{4}{10} =$

33). $\frac{3}{9} \div 2 =$

34). $16 \div \frac{4}{5} =$

35). $9 \div \frac{4}{8} =$

36). $10 \div \frac{3}{9} =$

37). $19 \div \frac{7}{9} =$

38). $4 \div \frac{1}{4} =$

39). $5 \div \frac{5}{6} =$

40). $20 \div \frac{8}{10} =$

ANSWER

1). $\dfrac{1}{2} \div 6 = \dfrac{1}{12}$

2). $\dfrac{1}{4} \div 5 = \dfrac{1}{20}$

3). $\dfrac{2}{3} \div 4 = \dfrac{2}{12}$

4). $\dfrac{2}{5} \div 2 = \dfrac{2}{10}$

5). $\dfrac{1}{5} \div 10 = \dfrac{1}{50}$

6). $\dfrac{5}{10} \div 7 = \dfrac{5}{70}$

7). $\dfrac{2}{4} \div 7 = \dfrac{2}{28}$

8). $\dfrac{2}{4} \div 5 = \dfrac{2}{20}$

9). $7 \div \dfrac{1}{2} = \dfrac{14}{1}$

10). $7 \div \dfrac{2}{4} = \dfrac{28}{2}$

11). $6 \div \dfrac{1}{4} = \dfrac{24}{1}$

12). $5 \div \dfrac{2}{4} = \dfrac{20}{2}$

13). $\dfrac{3}{5} \div 6 = \dfrac{3}{30}$

14). $\dfrac{1}{2} \div 6 = \dfrac{1}{12}$

15). $\dfrac{2}{4} \div 7 = \dfrac{2}{28}$

16). $\dfrac{4}{10} \div 7 = \dfrac{4}{70}$

17). $2 \div \dfrac{1}{5} = \dfrac{10}{1}$

18). $5 \div \dfrac{1}{2} = \dfrac{10}{1}$

19). $10 \div \dfrac{2}{10} = \dfrac{100}{2}$

20). $4 \div \dfrac{1}{2} = \dfrac{8}{1}$

21). $11 \div \dfrac{5}{8} = \dfrac{88}{5}$

22). $2 \div \dfrac{3}{8} = \dfrac{16}{3}$

23). $\dfrac{2}{9} \div 17 = \dfrac{2}{153}$

24). $10 \div \dfrac{4}{6} = \dfrac{60}{4}$

25). $\dfrac{4}{7} \div 2 = \dfrac{4}{14}$

26). $10 \div \dfrac{1}{8} = \dfrac{80}{1}$

27). $\dfrac{1}{6} \div 13 = \dfrac{1}{78}$

28). $\dfrac{1}{4} \div 2 = \dfrac{1}{8}$

29). $\dfrac{2}{5} \div 13 = \dfrac{2}{65}$

30). $13 \div \dfrac{3}{6} = \dfrac{78}{3}$

31). $\dfrac{1}{3} \div 20 = \dfrac{1}{60}$

32). $2 \div \dfrac{4}{10} = \dfrac{20}{4}$

33). $\dfrac{3}{9} \div 2 = \dfrac{3}{18}$

34). $16 \div \dfrac{4}{5} = \dfrac{80}{4}$

35). $9 \div \dfrac{4}{8} = \dfrac{72}{4}$

36). $10 \div \dfrac{3}{9} = \dfrac{90}{3}$

37). $19 \div \dfrac{7}{9} = \dfrac{171}{7}$

38). $4 \div \dfrac{1}{4} = \dfrac{16}{1}$

39). $5 \div \dfrac{5}{6} = \dfrac{30}{5}$

40). $20 \div \dfrac{8}{10} = \dfrac{200}{8}$